Changing Course

Changing Course
A Global Business
Perspective on Development
and the Environment

Stephan Schmidheiny
with the Business Council
for Sustainable Development

The MIT Press
Cambridge, Massachusetts
London, England

© 1992 Massachusetts Institute of Technology

This book was set in Palatino by The MIT Press and printed and bound in the United States of America.

The text of this book has been printed on recycled paper.

Library of Congress Cataloging-in-Publication Data

Schmidheiny, Stephan, 1947–
    Changing Course : a global business perspective on development and the environment / Stephan Schmidheiny with the Business Council for Sustainable Development.
        p. cm.
    Includes bibliographical references and index.
    ISBN 0-262-19318-3 (hc). — ISBN 0-262-69153-1 (pbk.)
    1. Economic development—Environmental aspects—case studies.
    2. Environmental policy—Costs—Case studies.   3. International business enterprises—Case studies.   4. Social responsibility of business—Case studies.   I. Title.
HD75.6.S35   1992
363.7'05765—dc20   92-6457
                                                                      CIP

Contents

To waste, to destroy, our natural resources, to skin and exhaust the land instead of using it so as to increase its usefulness, will result in undermining in the days of our children the very prosperity which we ought by right to hand down to them amplified and developed.

—Theodore Roosevelt
Message to Congress
December 3, 1907

Declaration of the Business Council for Sustainable Development

Business will play a vital role in the future health of this planet. As business leaders, we are committed to sustainable development, to meeting the needs of the present without compromising the welfare of future generations.

This concept recognizes that economic growth and environmental protection are inextricably linked, and that the quality of present and future life rests on meeting basic human needs without destroying the environment on which all life depends.

New forms of cooperation between government, business, and society are required to achieve this goal.

Economic growth in all parts of the world is essential to improve the livelihoods of the poor, to sustain growing populations, and eventually to stabilize population levels. New technologies will be needed to permit growth while using energy and other resources more efficiently and producing less pollution.

Open and competitive markets, both within and between nations, foster innovation and efficiency and provide opportunities for all to improve their living conditions. But such markets must give the right signals; the prices of goods and services must increasingly recognize and reflect the environmental costs of their production, use, recycling, and disposal. This is fundamental, and is best achieved by a synthesis of economic instruments designed to correct distortions and encourage innovation and continuous improvement, regulatory standards to direct performance, and voluntary initiatives by the private sector.

The policy mixes adopted by individual nations will be tailored to local circumstances. But new regulations and economic instruments must be harmonized among trading partners, while recognizing that levels and conditions of development vary, resulting in different needs and abilities. Governments should phase in changes over a reasonable period of time to allow for realistic planning and investment cycles.

Capital markets will advance sustainable development only if they recognize, value, and encourage long-term investments and savings, and if they are based on appropriate information to guide those investments.

Trade policies and practices should be open, offering opportunities to all nations. Open trade leads to the most efficient use of resources and to the development of economies. International environmental concerns should be dealt with through international agreements, not by unilateral trade barriers.

The world is moving toward deregulation, private initiatives, and global markets. This requires corporations to assume more social, economic, and environmental responsibility in defining their roles. We must expand our concept of those who have a stake in our operations to include not only employees and shareholders but also suppliers, customers, neighbors, citizens' groups, and others. Appropriate communication with these stakeholders will help us to refine continually our visions, strategies, and actions.

Progress toward sustainable development makes good business sense because it can create competitive advantages and new opportunities. But it requires far-reaching shifts in corporate attitudes and new ways of doing business. To move from vision to reality demands strong leadership from the top, sustained commitment throughout the organization, and an ability to translate challenge into opportunities. Firms must draw up clear plans of action and monitor progress closely.

Sustainability demands that we pay attention to the entire life cycles of our products and to the specific and changing needs of our customers.

Corporations that achieve ever more efficiency while preventing pollution through good housekeeping, materials substitution, cleaner technologies, and cleaner products and that strive for more efficient use and recovery of resources can be called "eco-efficient."

Long-term business-to-business partnerships and direct investment provide excellent opportunities to transfer the technology needed for sustainable development from those who have it to those who require it. This new concept of "technology cooperation" relies principally on private initiatives, but it can be greatly enhanced by support from governments and institutions engaged in overseas development work.

Farming and forestry, the businesses that sustain the livelihoods of almost half of the world's population, are often influenced by market signals working against efficient resource use. Distorting farm subsidies should be removed to reflect the full costs of renewable resources. Farmers need access to clear property rights. Governments should improve the management of forests and water resources; this can often be achieved by providing the right market signals and regulations and by encouraging private ownership.

Many countries, both industrial and developing, could make much better use of the creative forces of local and international entrepreneurship by providing open and accessible markets, more streamlined regulatory systems with clear and equitably enforced rules, sound and transparent financial and legal systems, and efficient administration.

We cannot be absolutely sure of the extent of change needed in any area to meet the requirements of future generations. Human history is that of expanded supplies of renewable resources, substitution for limited ones, and ever greater efficiency in their use. We must move faster in these directions, assessing and adjusting as we learn more. This process will require substantial efforts in education and training, to increase awareness and encourage changes in life-styles toward more sustainable forms of consumption.

A clear vision of a sustainable future mobilizes human energies to make the necessary changes, breaking out of familiar and established patterns. As leaders from all parts of society join forces in translating the vision into action, inertia is overcome and cooperation replaces confrontation.

We members of the BCSD commit ourselves to promoting this new partnership in changing course toward our common future.

Stephan Schmidheiny, Chairman of BCSD
Chairman
UNOTEC
Switzerland

Torvild Aakvaag   Naim Abou-Taleb
Vice-Chairman   Chairman &
Norsk Hydro A.S.   Managing Director
Norway   Mohandes Bank
                                     Egypt

Antonia Ax:son Johnson
Chairman
Axel Johnson AB
Sweden

Samuel C. Johnson
Chairman
S.C. Johnson & Son, Inc.
United States

Saburo Kawai
Vice Chairman & President
Keizai Doyukai
Japan

Jiro Kawake
Chairman
Oji Paper Co., Ltd.
Japan

Alex Krauer
Präsident des Verwaltungsrates
Ciba-Geigy AG
Switzerland

H. H., The Otunba Ayora,
(Mrs.) Bola Kuforiji-Olubi (M.O.N.)
Group Executive Chairman
BEWAC plc
Nigeria

Yutaka Kume
President
Nissan Motor Co. Ltd.
Japan

J.M.K. Martin Laing
Chairman
John Laing plc
United Kingdom

Erling S. Lorentzen
Chairman
Aracruz Celulose S.A.
Brazil

Ken F. McCready
President &
Chief Executive Officer
TransAlta Utilities Corp.
Canada

Akira Miki
Chairman
Nippon Steel Corporation
Japan

Jérôme Monod
Président Directeur Général
Lyonnaise des Eaux-Dumez
France

Shinroku Morohashi
President
Mitsubishi Corporation
Japan

Y.A.M. Tunku Naquiyuddin
 ibni Tuanku Ja'afar
Chairman
Antah Holdings Berhad
Malaysia

Philip Ndegwa
Chairman
First Chartered Securities Ltd.
Kenya

Paul H. O'Neill
Chairman &
Chief Executive Officer
ALCOA
United States

James Onobiono
Président Directeur Général
Compagnie Financière et
Industrielle CFI (S.A.)
Cameroon

Anand Panyarachun
Prime Minister of Thailand
Former Chairman of Saha-Union
Corp. Ltd.
Thailand

Frank Popoff
President &
Chief Executive Officer
The Dow Chemical Company
United States

Fernando Romero
Chairman
BHN Multibanco S.A.
Inversiones Bolivianas S.A.
Bolivia

William D. Ruckelshaus
Chairman of the Board
Chief Executive Officer
Browning-Ferris Industries
United States

Anthony Salim
President & CEO
Salim Group
Indonesia

Elisabeth Salina Amorini
Présidente du Conseil
d'Administration SGS
Société Générale de Surveillance
Holding S.A.
Switzerland

Helmut Sihler
Vorsitzender der
Geschäftsführung
HENKEL KGaA
Germany

Paul G. Stern
Chairman & CEO
Northern Telecom Ltd.
Canada

Ratan N. Tata
Chairman
TATA Industries Ltd.
India

Lodewijk C. van Wachem
Senior Managing Director
The Royal Dutch/Shell Group
The Netherlands/
United Kingdom

Sir Bruce Watson
Chairman
Mount Isa Mines Pty Ltd.
Australia

Edgar S. Woolard
Chairman of the Board
E.I. du Pont de Nemours
and Company
United States

Toshiaki Yamaguchi
President
Tosoh Corporation
Japan

Federico Zorraquin
President
S. A. Garovaglio y Zorraquin
Argentina

Preface

In mid-1990, Maurice Strong, secretary general of the 1992 U.N. Conference on Environment and Development, asked me to serve as his principal advisor for business and industry, to present a global business perspective on sustainable development and to stimulate the interest and involvement of the international business community.

I did not want to remain a lone voice. So I invited some 50 business leaders to become members of the Business Council for Sustainable Development (BCSD). Almost everyone I approached was prepared to join, which is an encouraging indication of the growing interest of business leaders in environment and development issues.

As we began to work together, I was impressed by the degree of interest they showed in involving themselves in complex issues that have traditionally been seen to belong to the realm of government, aid agencies, and environmental groups. This attitude clearly contrasts with any perception of business leaders as being concerned only with things of immediate relevance to their companies.

Our members took the trouble to participate actively in the preparatory process of the U.N. conference, joining with other parts of society to organize its agenda. They have taken a global, long-term perspective—looking beyond the immediate interests of themselves and their corporations, and beyond their own terms of office to the needs of future generations. And they have sought to transcend the boundaries of their own personal education, experience, and authority. They have not shied away from controversial topics characterized by incomplete or contradictory knowledge and understanding, conflicts of interests, or diverging personal opinions and preferences.

I would like to express my sincere appreciation for their entrepreneurial spirit, which led them to participate so fully and unselfishly in this demanding work.

My fellow members and I wanted very much to include in the BCSD representatives from Eastern Europe, the former Soviet Union, and China. This challenge proved to be beyond us. A shared concept of "business" allows people in Sweden, Australia, and Kenya to deal with one another. The absence of a "business sector" in the planned economies, as well as the complex problems emerging in the change from state-planned to market-driven systems, presented obstacles the Council was not able to overcome.

In early 1991, we selected a list of issues that we saw as necessary components of our report. At our first plenary meeting in The Hague in April 1991, we refined the list and began to divide into task forces to study and report on these issues. These groups were composed of members, their associates, outside experts, and members of the BCSD staff. BCSD members from developing countries and those from the industrial world agreed that the realities and needs of developing nations should permeate our book.

Each Council member delegated an associate as a representative to the more frequent Liaison Group meetings, which served to keep the Council members apprised of and engaged in the work of the Council secretariat and all the task forces. This group held three active and lively plenary meetings to review and discuss drafts of the various chapters.

But the BCSD is more than a product—this book. It is also a process. It has organized some 50 conferences, symposia, and issue workshops in more than 20 countries. Members from Africa, Asia, and Latin America were particularly active in their regions, offering their own experiences and viewpoints and drawing many of their peers into the process. At this writing, we expect that groups organized through the BCSD will produce their own reports on business and sustainable development in the world's three developing regions. The wide and continuing process may, in the long run, have a far greater impact than this product.

The fast-growing interest in sustainable development in the poorer parts of the world proves that people here understand that economic development can only happen in a healthy environment.

There is an inescapable logic in the concept of sustainable development. So it is hardly surprising that looking at core issues from a business point of view led the Council to many conclusions similar to those of other groups, such as economists and political leaders. But we also produced new perspectives that proved helpful to the planners of the U.N. Conference on Environment and Development, such as our discussion of technology cooperation.

Despite different national, cultural, and business backgrounds, we also managed to agree among ourselves. We did not agree on every detail and every word. That was never our intention; there was not enough time, and such unanimity would have required turning 50 hard-pressed business leaders into an editorial drafting committee.

But we have produced ideas and concepts that all have generally endorsed, and all have agreed to have their names listed with the BCSD Declaration. We sign on as individuals, but we realize our endorsement means that we will be closely watched and judged as we deploy our best efforts in our respective offices to comply with the program presented in this book.

A search for consensus can often produce a very low common denominator. I trust that our readers will agree that this is not the case in this book. We realize that many theories of sustainable development go far beyond the recommendations of our report. Indeed, the practices of some companies already go beyond our counsel. Yet this is the first time that an important group of business leaders has looked at these issues from a global perspective and reached major agreements on the need for an integrated approach in confronting the challenges of economic development and the environment. We have outlined a change of course for business that can have far-reaching consequences for most business people, including the BCSD members.

Our Perspective

The BCSD speaks not for global business but as a small group of business leaders, by definition representing a small minority. We claim no legitimacy beyond our collective wisdom and that of the many people who have worked on this report.

Yet *Changing Course* does offer a global perspective, because sustainable development itself is a global vision. Such a perspective must by nature be general; it cannot be specific about countries, business sectors, or individual corporations. We have offered in our text specific examples from the real world to demonstrate the practical relevance of our statements. We have also collected, with the help of many corporations around the world, short case studies to demonstrate that what we offer is not abstract but is already being done by many of the more successful companies.

The challenges of sustainable development are of a magnitude and complexity that far surpass the understanding of any individual or group. We entered this exercise in a spirit of humility, and have emerged even more humble.

Our work should be compared with the initial sketches of an architect who is concerned with the concept and the shape of a building, its functions, and its different relations with its environment. These sketches must be elaborated upon by many—measuring, calculating, and specifying—before they are converted to blueprints. Then the job of turning design into reality falls to many builders; we include ourselves in their number.

We call for a long-term view, for far-reaching changes, and for action. But we do not base our hopes for success on radical changes in human nature or on the creation of a utopia. We take humans the way we find them, the way we all are made, with all our strengths and weaknesses. We base the conclusions in our report on the facts and our own experiences of the real world. We believe that given the will and understanding, our proposals can eventually become part of practical reality.

The title *Changing Course* was chosen with some care. While the basic goal of business must remain economic growth, as long as world population continues to grow rapidly and mass poverty remains widespread, we are recommending a different course toward that goal. There will be changes in direction and changes in the measurements of progress to include indicators of quality as well as quantity. Business is a large vessel; it will require great common effort and planning to overcome the inertia of the present destructive course, and to create a new momentum toward sustainable development.

A Personal Note

Although for practical reasons my name stands as that of the author of *Changing Course*, credit for most of the content of this book belongs to others. The names of those who have contributed a great deal, aside from the BCSD members, appear in Appendix 2. But I would like to express here my special thanks to BCSD executive director Hugh Faulkner, who managed our entire process with great skill and a rare mix of vision, commitment, and good will, and to our editorial advisor, Lloyd Timberlake, who helped us find words for our findings and conclusions.

Working together toward a common goal with many people of different backgrounds in many parts of the world has been a most rewarding experience. From the beginning, I have seen this not as a philanthropic effort, but rather as an investment in my own education and in the future of my children's business. As happens with so many investment projects, it came in over budget: I promised Maurice Strong a third of my time over 20 months, and I have in fact found more than half of my time taken up by BCSD work.

Looking back—and looking ahead—I have every reason to be satisfied that I did not spare any effort. I feel sure that the investment will reap substantial dividends.

I found the combination of business and environment concerns appropriate. Conservation of the environment and successful business development should be opposite sides of the same coin—the coin being the measure of the progress of human civilization. The degree to which these two halves can be joined in the world of human activity, and the speed of this process, will determine the rate at which sustainable development will turn from a vision into reality.

Stephan Schmidheiny
January 1992

Changing Course

1   The Business of Sustainable Development

"With greater freedom for the market comes greater responsibility."

Gro Harlem Brundtland
Prime Minister, Norway

Running a company requires daily assessments of opportunities, risks, and trends. Corporate leaders who ignore economic, political, or social changes will lead their companies toward failure. So too will those who overreact to change and perceived risk.

Many global trends offer hope. Life expectancy, health care, and education have all improved dramatically in the second half of this century. World food production has stayed well ahead of population growth. Average per capita incomes have increased by the highest sustained rates ever. No shortage of raw materials looms in the foreseeable future. Given the right technology, the planet's soils can supply more than the basic food needs of much larger populations.

But neither business nor any other leaders can afford to see only the positive, especially when these optimistic signs are based on averages that mask alarming departures from the norm. Several other linked global trends demand any thinking person's attention. Each is replete with scientific uncertainty. To overreact to any of these would be dangerous, but to ignore any of them would be irresponsible.

First, the human population is growing extremely rapidly; according to the most optimistic estimates, an already crowded planet is likely to have to support twice as many people next century. Environmental ills have varying causes, but all are made worse by the pressure of human numbers.

Second, the last few decades have witnessed an accelerating consumption of natural resources—consumption that is often inefficient and ill planned. Resources that biologists call renewable are not being given time to renew. The bottom line is that the human species is living more off the planet's capital and less off its interest. This is bad business.

Third, both population growth and the wasteful consumption of resources play a role in the accelerating degradation of many parts of the environment. Productive areas are hardest hit. Agriculturally fruitful drylands are turning into desert; forests into poor pastures; freshwater wetlands into salty, dead soils; rich coral reefs into lifeless stretches of ocean.

Fourth, as ecosystems are degraded, the biological diversity and genetic resources they contain are lost. Many environmental trends are reversible; this loss is permanent.

Fifth, this overuse and misuse of resources is accompanied by the pollution of atmosphere, water, and soil—often with substances that persist for long periods. With a growing number of sources and forms of pollution, this process also appears to be accelerating. The most complex and potentially serious of these threats is a change in climate and in the stability of air circulation systems.

There are also alarming trends apparent in patterns of "development"; these, too, begin with population projections. More than 90 percent of population growth takes place in the developing world—that is, in poorer countries.[1] This means that when the present world population of more than 5 billion doubles next century, there will be an extra 4.5 billion people in nations where today it is hardest to secure jobs, food, safe homes, education, and health care.

Already, population growth means that the number of people who belong to the underclass—those unable to secure such basics of life as adequate food, shelter, clothing, health care, and education—is rising yearly in much of the developing world. In the mid-1980s more than a billion people on the planet, almost a third of the population of the developing world, were trying to survive on an income equivalent to about $1 per day.[2]

Poverty, rapid population growth, and the deterioration of natural resources often occur in the same regions, creating a huge imbalance between the quarter of the planet living in rich, industrial nations and the three quarters residing in developing nations. The national income of Japan's 120 million people is about to overtake the combined incomes of the 3.8 billion people in the developing world.[3] The industrial nations have generally cut their aid to the developing world. And because of debt servicing and repayment and reduced foreign investment, total capital flows were reversed in the second half of the 1980s, with money flowing from poor to rich.

These two sets of alarming trends—environment and development—cannot be separated. Economic growth in most of the developing world will depend for some time on agricultural production, so a reduction in the productivity in ecosystems tends to mean declining farm production and loss of revenue. But increasing populations in stagnant economies mean that more people are directly relying on the environmental resource base for a living. Environmental and economic decline are in many areas an inseparable part of the same downward spiral.

The resulting global structural challenge was concisely summed up by Maurice Strong, Secretary General of the U.N. Conference on Environment and Development in Brazil in June 1992: "The gross imbalances that have been created by concentration of economic growth in the industrial countries and population growth in developing countries is at the centre of the current dilemma. Redressing these imbalances will be the key to the future security of our planet—in environmental and economic as well as traditional security terms. This will require fundamental changes both in our economic behavior and our international relations."[4]

Clearly action is required. But which actions, and when, given the huge uncertainties involved? This is the sort of issue that business copes with daily. Corporate leaders are used to examining uncertain, negative trends, making decisions, and then taking action, adjusting, and incurring costs to prevent damage. Insurance is just one example. There are costs involved, but these are costs the rational are willing to bear and costs the responsible do not regret, even if things turn out not to have been as bad as they once seemed. We can hope for the best, but the "precautionary principle" remains the best practice in business as well as in other aspects of life.

This principle was agreed to at the World Industry Conference on Environmental Management in 1984 and at the 1989 Paris summit of the leaders of the seven richest industrial nations (the G7). It was strengthened in the Ministerial Declaration of the 1990 U.N. Economic Commission for Europe meeting in Bergen: "In order to achieve sustainable development, policies must be based on the precautionary principle. Environmental measures must anticipate, prevent and attack the causes of environmental degradation. Where there are threats of serious or irreversible damage, lack of full scientific certainty should not be used as a reason for postponing measures to prevent environmental degradation."[5]

Yet risk and uncertainty are usually accompanied by new opportunities, and business has long been adept at seizing such chances. For example, using energy more efficiently, and thus reducing global and local pollution, decreases costs and increases competitiveness. Many of what politicians and economists call "no regrets policies"—actions that make sense no matter what the real threat of global warming—are from a business point of view opportunities and good investments; examples include improving energy efficiency and developing new energy sources, new drought-resistant crops, and new resource management techniques.

There will be many opportunities for business. Having researched various aspects of competitive advantage among nations, Harvard Business School Professor Michael Porter reported: "I found that the nations with the most rigorous [environmental standards] requirements often lead in exports of affected products....The strongest proof that environmental protection does not hamper competitiveness is the economic performance of nations with the strictest laws." He mentions the successes of Japan and Germany, as well as that of the United States in sectors actually subject to the greatest environmental costs: chemicals, plastics, and paints.[6]

Viewing environmental threats from a business perspective can help guide both governments and companies toward plausible policies that offer protection from disaster while making the best of the challenges.

This book is about the steps required of business, and of the governments that set the frameworks for industry, to ensure that humans and all other species continue to occupy a safe and bountiful planet.

Sustainable Development

During the first great wave of environmental concern in the late 1960s and early 1970s, most of the problems seemed local: the products of individual pipes and smokestacks. The answers appeared to lie in regulating these pollution sources.

When the environment reemerged on the political agenda in the 1980s, the main concerns had become international: acid rain, depletion of the ozone layer, and global warming. Analysts sought causes not in pipes

> *"This is the moment of truth for Western Europe and the industrialized world. Will we prove to be strong enough and refrain from part of our own consumption in order to secure a peaceful and democratic development in Eastern Europe? Will we be able to give hope to all the poor, who for so long have been oppressed by an inhuman system and denied economic development as well as an acceptable environment?"*
>
> Percy Barnevik
> Chief Executive Officer
> ABB Asea Brown Boveri Ltd.

and stacks but in the nature of human activities. One report after another concluded that much of what we do, many of our attempts to make "progress," are simply unsustainable. We cannot continue in our present methods of using energy, managing forests, farming, protecting plant and animal species, managing urban growth, and producing industrial goods. We certainly cannot continue to reproduce our own species at the present rate.

Energy provides a striking example of present unsustainability. Most energy today is produced from fossil fuels: coal, oil, and gas. In the mid-1980s, the world was burning the equivalent of 10 billion metric tons of coal per year, with people in industrial nations using much more than those of the developing world. At these rates, by 2025 the expected global population of more than 8 billion would be using the equivalent of 14 billion metric tons of coal. But if all the world used energy at industrial-country levels, by 2025 the equivalent of 55 billion metric tons would be burned. Present levels of fossil fuel use may be warming the globe; a more than fivefold increase is unthinkable. Fossil fuels must be used more efficiently while alternatives are being developed if economic development is to be achieved without radically changing the global climate.[7]

Given such widespread evidence of unsustainability, it is not surprising that the concept of "sustainable development" has come to dominate the environment/development debate. In 1987, the World Commission on Environment and Development, appointed three years earlier by the U.N. General Assembly and headed by Norwegian Prime Minister Gro Harlem Brundtland, made sustainable development the theme of its entire report, *Our Common Future*. It defined the concept simply as a form of development or progress that "meets the needs of the present without compromising the ability of future generations to meet their own needs."[8]

The phrase itself can be misleading, as the word development might suggest that it is a chore for "developing" nations only. But development is more than growth, or quantitative change. It is primarily a change in quality. More than a decade ago, the influential World Conservation Strategy, compiled by the United Nations and organizations representing governments and private bodies, defined development as "the modification of the biosphere and the application of human, financial, living and non-living resources to satisfy human needs and improve the quality of human life."[9]

Thus all nations are, or would wish to be, developing. And sustainable development will require the greatest changes in the wealthiest nations, which consume the most resources, release the most pollution, and have the greatest capacity to make the necessary changes. These nations must also respond to the criticism from many leaders in the poor parts of the world that industrial countries risk reversing the relationship between production and the satisfaction of needs. They charge that increased production in wealthy nations no longer serves primarily to satisfy needs; rather, the creation of needs serves to increase production.

The idea that much of what humanity does in the name of progress is unsustainable and must be changed has gained rapid acceptance. In 1987, the U.N. General Assembly passed a resolution adopting the World Commission's report as a guide for future U.N. operations, and commending it to governments. Since then, many governments have tried to bend their policies to its recommendations. The July 1989 G7 summit called for "the early adoption, worldwide, of policies based on sustainable development."[10]

Business has also taken up the challenge, at the international, national, and sectoral levels. The International Chamber of Commerce drafted a "Business Charter for Sustainable Development," which was launched in April 1991 at the Second World Industry Conference on Environmental Management. The Charter, endorsed by 600 firms worldwide by early 1992, encourages companies to "commit themselves to improving their environmental performance in accordance with these [the Charter's] 16 Principles, to having in place management practices to effect such improvement, to measuring their progress, and to reporting this progress as appropriate internally and externally."[11]

The senior business group in Japan, the Keidanren, adopted an Environmental Charter in 1991 that sets out codes of behavior toward the environment.[12] Malaysia has established a corporate environmental policy that calls on companies "to give benefit to society; this entails...that any adverse effects on the environment are reduced to a practicable minimum."[13] The Confederation of Indian Industry has also urged an "Environment Code for Industry" upon its members.[14]

Chemical industry associations in several countries have agreed to a Responsible Care program to promote continuous improvement in environmental health and safety. Begun in Canada and taken up in the United States, Australia, and many European countries, the scheme encourages associations to draft codes of conduct in many areas of operations, and it recommends that large companies help smaller ones with environmental and safety improvements.[15]

Sustainable development will obviously require more than pollution prevention and tinkering with environmental regulations. Given that ordinary people—consumers, business people, farmers—are the real day-to-day environmental decision makers, it requires political and economic systems based on the effective participation of all members of society in decision making. It requires that environmental considerations become a part of the decision-making processes of all government agencies, all business enterprises, and in fact all people. It requires levels of international cooperation never before achieved, not least in agreeing to and enforcing treaties to manage global commons such as the atmosphere and oceans. It requires, beyond immediate environmental concerns, an end to the "arms culture" as a method of achieving security, and new definitions of security that include environmental threats.

Recently the nations of the world seem to have begun to move, albeit slowly, in these directions. Environmental concern has gradually begun to infuse all areas of decision making. Democracy has become a more prevalent form of government throughout the developing world, Eastern Europe, and the former Soviet Union. The Montreal protocol on the ozone layer and an emerging treaty on the atmosphere suggest that nations may be able to cooperate along the harder paths toward a cleaner global environment. Definitions of security are changing, and the end of the cold war may free resources for work on environmental security.

Will these changes last, and are they happening fast enough?

The Growth Controversy

Perhaps the most controversial conclusion of the World Commission was that sustainable development requires rapid economic growth. This assumption is based on the reality that growing populations and poor populations require goods and services to meet essential needs and that "meeting essential needs depends in part on achieving full growth potential, and sustainable development clearly requires economic growth in places where such needs are not being met. Elsewhere, it can be consistent with economic growth, provided the content of growth reflects the broad principles of sustainability and non-exploitation of others."[16]

There are critics who argue that the limits to economic growth have already been reached and there must be "zero growth" from now on. Some of them do not explain how zero growth will meet the needs of a planet with more than 10 billion people. But others who feel that the environmental limits to growth have been reached, such as Robert Goodland and Herman Daly of the World Bank's Environment Department, argue that "development by the rich must be used to free resources...for growth and development so urgently needed by the poor. Large-scale transfers to the poorer countries also will be required."[17]

The World Commission itself noted that growth alone is not enough, as high levels of productivity and widespread poverty can coexist and can endanger the environment. So sustainable development requires societies to meet human needs both by increasing environmentally sustainable production and "by ensuring equitable opportunities for all."[18]

The Commissioners argued that growth was limited at present by both the nature of technologies and the nature of social organization. For example, humanity will eventually be using much more energy than today, but an increasing part will have to come from sources other than fossil fuels. And if societies remain organized so that many people remain impoverished despite sustained economic growth, then this poverty will both degrade environmental resources and eventually act as a brake on growth.

The Business Challenge

The requirement for clean, equitable economic growth remains the biggest single difficulty within the larger challenge of sustainable development.

Proving that such growth is possible is certainly the greatest test for business and industry, which must devise strategies to maximize added value while minimizing resource and energy use. Given the large technological and productive capacity of business, any progress toward sustainable development requires its active leadership.

Open, prospering markets are a powerful force for creating equity of opportunity among nations and people. Yet for there to be equal opportunity, there must first be opportunity itself. Open, competitive markets create the most opportunities for the most people. It is often the nations where markets most closely approach the ideal of "free," open, and competitive that have the least poverty and the greatest opportunity to escape from that poverty.

The World Commission listed as the first prerequisite for sustainable development a political system in which people can effectively participate in decision making. But freedom to participate in political decisions and freedom to participate in markets are inseparable over the long run. The citizens of Central Europe, having achieved political freedom, are now building market freedom. The Asian nations that achieved thriving market economies under authoritarian regimes are now moving toward more democratic governments.

Yet no market can be called "free" in which the decisions of a few can cause misuse of resources and pollution that threaten the present and future of the many. Today, for instance, the earth's atmosphere is providing the valuable service of acting as a dump for pollutants; those enjoying this service rarely pay a reasonable price for it.

"Eco-efficiency"

The present limits to growth are not so much those imposed by resources, such as oil and other minerals, as was argued by the 1972 Club of Rome report *The Limits to Growth*.[19] In many cases they arise more from a scarcity of "sinks," or systems that can safely absorb wastes. The atmo-

> *"We believe a business cannot continue to exist without the trust and respect of society for its environmental performance."*
>
> Shinroku Morohashi
> President
> Mitsubishi Corporation

sphere, many bodies of water, and large areas of soil are reaching their own absorptive limits as regards wastes of all kinds.

Business has begun to respond to this truth. It is moving from a position of limiting pollution and cleaning up waste to comply with government regulations toward one of avoiding pollution and waste both in the interests of corporate citizenship and of being more efficient and competitive. The economies of the industrial countries have grown while the resources and energy needed to produce each unit of growth have declined. Chemical companies in industrial nations have doubled output since 1970 while more than halving energy consumption per unit of production.[20]

Industry is moving toward "demanufacturing" and "remanufacturing"—that is, recycling the materials in their products and thus limiting the use of raw materials and of energy to convert those raw materials. (See chapter 7.) That this is technically feasible is encouraging; that it can be done profitably is more encouraging. It is the more competitive and successful companies that are at the forefront of what we call "eco-efficiency."

But eco-efficiency is not achieved by technological change alone. It is achieved only by profound changes in the goals and assumptions that drive corporate activities, and change in the daily practices and tools used to reach them. This means a break with business-as-usual mentalities and conventional wisdom that sidelines environmental and human concerns.

A growing number of leading companies are adopting and publicly committing themselves to sustainable development strategies. They are expanding their concepts of who has a stake in their operations beyond employees and stockholders to include neighbors, public interest groups (including environmental organizations), customers, suppliers, governments, and the general public. They are communicating more openly with these new stakeholders. They are coming to realize that "the degree to which a company is viewed as being a positive or negative participant in solving sustainability issues will determine, to a very great degree, their long-term business viability," in the words of Ben Woodhouse, director of Global Environmental Issues at Dow Chemical.

The Challenge of Time

As the World Commission noted, sustainable development requires forms of progress that meet the needs of the present without compromising the ability of future generations to meet their own needs. In the late twentieth century, we are failing in the first clause of that definition by not meeting the basic needs of more than 1 billion people. We have not even begun to come to grips with the second clause: the needs of future generations. Some argue that we have no responsibility for the future, as we cannot know its needs. This is partly true. But it takes no great leap of reason to assume that our offspring will require breathable air, drinkable water, productive soils and oceans, a predictable climate, and abundant plant and animal species on the planet they will share.

Yet it is a hard thing to demand of political leaders, especially those who rely on the votes of the living to achieve and remain in high office, that they ask those alive today to bear costs for the sake of those not yet born, and not yet voting. It is equally hard to ask anyone in business, providing goods and services to the living, to change their ways for the sake of those not yet born, and not yet acting in the marketplace. The painful truth is that the present is a relatively comfortable place for those who have reached positions of mainstream political or business leadership.

This is the crux of the problem of sustainable development, and perhaps the main reason why there has been great acceptance of it in principle, but less concrete actions to put it into practice: many of those with the power to effect the necessary changes have the least motivation to alter the status quo that gave them that power.

When politicians, industrialists, and environmentalists run out of practical advice, they often take refuge in appeals for a new vision, new values, a new commitment, and a new ethic. Such calls often ring hollow and rhetorical. But given that sustainable development requires a practical concern for the needs of people in the future, then it does ultimately require a new shared vision and a collective ethic based on equality of opportunity not only among people and nations, but also between this generation and those to come. Sustainability will require new technology, new approaches to trade to spread the technology and the goods necessary for survival, and new ways of meeting needs through markets. Business leadership will be required, and expected, in all these areas.

However, sustainable development will ultimately be achieved only through cooperation among people and all their various organizations, including businesses and governments. And leaders elected to decision-making and executive offices retain a fundamental obligation to inform and educate their constituencies about the urgent necessity and the reasons for changing course.

We believe that the best aspects of the human propensity to buy, sell, and produce can be an engine of change. Business has helped to create much of what is valuable in the world today. It will play its part in ensuring the planet's future.

Shaping the Future

The inevitable process of change toward sustainable forms of development will determine the future course of human civilization and shape our life-styles and thereby the way we do business. Yet many business leaders have so far been relatively passive in dealing with these issues.

Perhaps this ultimately has to do with the way we each react to the dimension of time. Those who have little interest in the future of nature and humanity will have little regard for the sustainability of their actions and will not be concerned to grapple with and understand the challenges facing us all. Those who do care about society and its progress have learned to understand that business never operates in a vacuum. It interacts at many levels with society, and society is now entering a period of rapid and fundamental changes.

Business has developed remarkable skills in market intelligence to spot and to a certain extent predict changing demand patterns. It must also construct a system of "social intelligence" to spot, understand, and interpret signals of change in development patterns. Those who are quickest to receive and act on such signals will have a great advantage over competitors who react only when changes in society become apparent in the form of changed consumer habits.

The environmental challenge has grown from local pollution to global threats and choices. The business challenge has likewise grown—from relatively simple technical fixes and additional costs to a corporatewide collection of threats, choices, and opportunities that are of central importance in separating tomorrow's winners from tomorrow's losers. Corporate leaders must take this into account when designing strategic plans of business and deciding the priorities of their own work.

Sustainable development is also about redefining the rules of the economic game in order to move from a situation of wasteful consumption and pollution to one of conservation, and from one of privilege and protectionism to one of fair and equitable chances open to all. Business leaders will want to participate in devising the rules of the new game, striving to make them simple, practical, and efficient.

No one can reasonably doubt that fundamental change is needed. This fact offers us two basic options: we can resist as long as possible, or we can join those shaping the future. *Changing Course* is an invitation and a challenge to business leaders to choose the more promising and more rewarding option of participation.

2   Pricing the Environment: Markets, Costs, and Instruments

"Bureaucratic socialism collapsed because it did not allow prices to tell the economic truth. Market economy may ruin the environment and ultimately itself if prices are not allowed to tell the ecological truth."

Ernst U. von Weizsäcker
Institut für Klima, Umwelt + Energie GmbH
(Institute for Climate, Environment, and Energy)

The cornerstone of sustainable development is a system of open, competitive markets in which prices are made to reflect the costs of environmental as well as other resources.

This assertion is the basis of much that follows in our book, and we hold it to be true for many reasons. Sustainable development requires the production of increasing amounts of goods and services to meet the needs of rapidly growing numbers of people. At the same time, the use of environmental resources must become more efficient, and production processes and consumption patterns must release less pollution. Given that production and pollution are influenced by the daily activities of billions of people, sustainable development cannot be secured efficiently by government decisions alone. Governments should instead provide the framework in which it can happen.

Open markets can motivate people toward sustainable development. When resources are priced properly, the pursuit of competitiveness encourages producers to minimize resource use. To the extent that pollution represents resources that have "escaped" from a production system, the concern for costs will also encourage producers to minimize pollution, especially when they pay to control it or are made liable for damage it causes. The competition inherent in open markets is the primary driving force for the creation of new technology. And new technology is needed to use resources more efficiently and further reduce pollution. Inasmuch as the development of technology also depends on economic growth, so, then, does sustainable development.

Open, competitive markets create jobs and opportunities, and thus are the most effective way of meeting people's needs. Accessible markets empower people, and offer the greatest opportunities to the poor. Free markets are also inseparable from the other forms of freedom sweeping the planet at the close of the twentieth century, not only in Eastern Europe but also in Latin America, Africa, and Asia. The barriers against citizens' effective participation in markets are a major reason for the failure of centrally planned economies in Eastern Europe and the slow, faltering development of many developing nations.

But the fall of communism does not represent the total victory of capitalism. It is merely the end of a system that, as practiced in Eastern Europe and in the Soviet Union, reflected neither economic nor environmental truths. This should encourage those of us who believe in the efficacy of the marketplace to eliminate its failures and weaknesses and to build on its strengths. Market economies must now rise to the challenge and prove that they can adequately reflect environmental truth and incorporate the goals of sustainable development.

Making Markets Work for the Environment

If markets really do encourage efficient resource use and decreases in pollution, then we must ask ourselves why the past record of industrialization is largely one of unsustainable resource use and high levels of pollution.

The primary answer is that markets have simply not efficiently reflected the costs of environmental degradation. They often fail to integrate environmental costs into economic decisions—at either the business or the government level. These costs are called "externalities." They traditionally do not enter into cost calculations other than through environmental regulations, many of which do not go far enough and which are unnecessarily burdensome to consumers and industry. For example, a coal-fired power plant that releases pollution plays a role in damaging human health, corroding the built environment, killing forests, and acidifying lakes. There is nothing theoretical about the costs of this damage. But they are spread throughout society and are external to the operations of the power plant.

For nations to factor these externalities into the costs of doing business is probably the most important correction necessary in current market systems. This basic defect can be mended by a variety of instruments. Indeed, many businesses and many governments have already started the process. Many existing environmental laws are a means of internalizing environmental costs. Ultimately, however, businesses and governments must undertake the process in an internationally harmonized manner and at least begin to move in the same direction, so that the global marketplace offers the same information, the same rules of competition, and therefore the same openness and opportunities for all participants.

"Full-Cost Pricing"

Including environmental costs in the accounts of business is often referred to as "full-cost pricing." This is a theoretical abstraction, an ideal. Like other powerful ideals, such as democracy and the concept of the "free market" itself, its approximation will vary under different conditions in different times and places.

Economists are working to establish the "full cost" of various types of pollution and other environmental damage. Although this work is useful, some costs cannot be quantified accurately. For example, in the case of global pollution problems such as ozone depletion, the full cost remains an abstraction because the price of destroying the planet's ozone layer cannot be accurately estimated. But the feared health impacts of ozone loss by themselves encouraged acts of control. The task of internalizing environmental costs must proceed using imperfect, existing knowledge and imperfect, available tools.

The lack of accuracy in determining the actual and future costs of pollution should not allow us to conclude that no price can be established at all. As individuals set prices for privately owned goods, society must establish through political processes prices for the use of goods held in common: waters, atmosphere, and so on. This work must be based both on the best available scientific evidence and on people's preferences and choices.

The market does not tell us where to go, but it provides the most efficient means of getting there. Therefore, society—through its political systems—will have to make value judgments, set long-term objectives, implement measures such as charges and taxes step by step, and make midcourse corrections based on experience and changing evidence. Thus in moving toward sustainable development, it may be sufficient merely to introduce environmental charges slowly but predictably. Infrastructure planning, technology development, cultural patterns, and consumers could then anticipate the price increase and react accordingly.

The basic equations behind full-cost pricing are simple. For production, whether industrial or agricultural, the full cost is the cost of production plus the cost of any environmental damage associated with it. For use of resources, such as minerals or forests, the full cost is the cost of extraction plus the cost of environmental damage. For example, clearing a forest might degrade environmental services provided by that forest, such as protecting a watershed.

Resource extraction, manufacturing, and energy use and consumption levels are all affected by cost considerations. What is extracted, produced, and consumed depends on price. The higher the price, the lower the demand. This basic economic principle applies to the environment as well as to commercial transactions. When there is no or only a small charge to either individuals or corporations for dumping waste into the environment, more waste will be dumped than would be the case if prices reflected the full cost of disposal.

Possibly the most important factor in an effective pursuit of sustainable development is "getting the price right." Unless prices for raw materials and products properly reflect the social costs, and unless prices can be assigned to air, water, and land resources that presently serve as cost-free receptacles for the waste products of society, resources will tend to be used inefficiently and environmental pollution will likely increase.

Many production, extraction, and distribution processes do not even pay the economic costs of the activity (land, labor, and capital costs), much less the added costs of environmental damage. This is because elements of these activities are subsidized. Subsidies that degrade the environment are not a market failure but a policy failure. In many industrial nations, some forms of energy and transport, water, and some agricultural products are on the market for a price less than their cost. In other cases, prices are kept far above economic costs, encouraging wasteful oversupply (as with farm surpluses) and environmentally damaging, overintensive methods of production. In many developing countries, energy, water, land, and agricultural inputs such as pesticides tend to be available at less than cost. The effect is wasteful overuse, resulting in increasing pollution and salinization of soils.

These subsidies have been introduced for a variety of political and social reasons over time, with little coordination of their impacts. Their beneficiaries—whether they be farmers, miners, corporations, or consumers—are all voters and have gotten used to the support they provide. Removing a subsidy is usually more difficult than introducing one. Because the process of eliminating subsidies creates adjustment problems, the financial resources currently used to support those that contribute to environmental problems should be redirected to financing the adjustment process of phasing them out.

The Polluter Pays Principle

Although the concept of internalizing costs is understood within the theory of economics, its realization becomes controversial in the real world. But one aspect of the debate has moved at least to the stage of international agreement.

As early as 1972 the members of the Organisation for Economic Co-operation and Development (OECD) agreed to the "Polluter Pays Principle" (PPP), which states that polluters should bear the full costs of any damage caused by the production of goods and services. This precept has gained wide recognition in both industrial and developing countries. Since the OECD policy allows any mechanism used to internalize costs to be counted toward meeting the PPP, much of the existing environmental regulation is consistent with the principle. Yet its implementation remains imprecise and somewhat ad hoc.

Although it still raises questions of what the real costs are, the point of departure of any attempt to internalize environmental costs must be this principle endorsed by all OECD members 20 years ago. In moving more systematically to the introduction of the Polluter Pays Principle, we urge governments to distinguish between easily identified costs and those impacts that pose complex questions of valuation. Progress on the former should not be delayed by debate on the latter. Where the economic impacts of this principle are significant and disruptive, a transition approach with identified end costs should be used. Governments also must clarify issues of the costs of past damage, which are not implicitly covered by the PPP.

It is often argued that it is not the "polluter," in the form of the producer, who pays, but that part of the cost is passed along to the consumer. But this was the point of the principle from its very inception. Though there are various types of pollution charges and they have various primary objectives, most result in higher product prices. Higher prices for more environmentally damaging products send a market signal to the consumer to seek a cleaner substitute. As consumers respond, so do producers. The Polluter Pays Principle is meant to affect the choices of consumers. Open, competitive markets ensure that industry reacts to changes in consumer demand swiftly and efficiently.

Inducing Change

Three basic mechanisms can be used to move business to internalize environmental costs, to pay for the costs of pollution, or to limit damage to the environment by other means:

• *Command and control:* These are basically government regulations, including performance standards for technologies and products, effluent and emission standards, and so on.

• *Self-regulation:* These are initiatives by corporations or sectors of industry to regulate themselves through standards, monitoring, pollution reduction targets, and the like.

• *Economic instruments:* These are efforts to alter the prices of resources and of goods and services in the marketplace via some form of government action that will affect the cost of production and/or consumption.

Command and Control

Traditionally, governments have used command-and-control regulations to achieve environmental objectives. "Performance standards" set a target—often for emissions—and allow companies flexibility in meeting it; "prescriptive standards" may prescribe the actual technology to be used, assuming that it will achieve the desired result. The former allow companies more scope for innovation and efficiency.

There are emerging concerns, however, shared by both governments and industry about the useful limits and constraints associated with the regulatory approach:

• Regulations are inflexible and often not the most cost-effective way of achieving change.

• Regulation requires adherence to specific requirements (such as air filters), as opposed to continuous improvement and continuous innovation, so it tends to perpetuate rather than improve the state of the art.

• The regulatory approach poses particular problems for developing countries, where the required administrative infrastructure to assure compliance may not be sufficient for the task.

Our view is that regulations have served a useful purpose and that there will continue to be a need for a basic regulatory framework in all countries. Command and control is particularly useful when there is a serious threat to health or safety or when pollution becomes especially dangerous once it exceeds a given level locally.

Self-Regulation

Self-regulation has achieved and will continue to achieve important improvements in the environmental impacts of business and industry. But for it to work and be credible, a clear framework of expectations and requirements must be negotiated between participating industries and government. Within such a framework, industry will be free to innovate and compete.

Various forces drive industry toward self-regulation:

• the threat of government regulation (which requires a government with a proven regulatory apparatus);

• solicited voluntary initiatives (akin to the threat of regulation, in which the regulatory body "encourages" a company to take certain environmental initiatives);

• required disclosure of environmental effects (this is based on a government regulation, but is relatively inexpensive for both governments and companies);

• public pressure or public esteem (industries operate with an implied contract with the public; loss of confidence cannot long be tolerated);

• peer pressure (leading companies are adopting sustainable development charters, which puts pressure on others to do likewise);

• a common view of common threats (industrialists are people too, and take seriously threats to the human environment and human well-being that ordinary citizens worry about); and

• consistency of behavior by multinational corporations (international media communications means that multinational corporations receive intense scrutiny; this provides both an opportunity and an obligation to use consistently high standards in all operations).[1]

Self-regulation may prove cheaper to society in general than either command-and-control regulations or economic instruments. Industry often holds the information on technologies and emissions that government needs to regulate effectively. Self-regulation thus to some extent avoids the expense of governments collecting that information, processing it into regulations, and then monitoring the effect. Obviously, there will still be information gathering and monitoring by governments, even in a system of extensive self-regulation, but it should be less adversarial and less extensive, and therefore cheaper.

Self-regulation does, however, have its drawbacks. It could lead to the creation of cartels and protectionism. And it can be frustrated by "free rider" companies using noncompliance to gain unfair competitive advantage. One response to this problem is simply to rely on market forces; many free riders will suffer the consequences of noncompliance because the ecologically sound behavior agreed to by the self-regulating companies is sound business behavior. Alternatively, complying companies and industry trade associations that foster self-regulation can exert pressure on free riders, or they can ask governments to introduce regulations based on the companies' or associations' own model.

Economic Instruments

The growing interest in the use of economic instruments stems from four needs: to provide continuous rewards and incentives for continuous improvements, to use markets more effectively in achieving environmental objectives, to find more cost-effective ways for both government

> *"While maintaining a market economy, the government needs to change the signals so that sustainable development becomes possible. Not everything can be achieved by regulations. It is important to bring about responsible behavior through incentives, such as taxes on energy consumption and clean water instead of on employment and the creation of wealth."*
>
> Eugenio Clariond Reyes
> Presidente Ejecutivo
> Grupo IMSA, S.A.

and industry to achieve these same objectives, and to move from pollution control to pollution prevention.

There is no clear, agreed definition of what constitutes an economic instrument, but all of them involve intervention by government in the marketplace through mechanisms such as pollution taxes and charges, tradable pollution permits and resource quotas, deposit-refund systems (as with glass bottles), performance bonds, resource saving credits, differential prices (as with unleaded versus leaded gasoline), special depreciation provisions, and the removal of subsidies and barriers to market activity.[2]

Many of these instruments raise revenue. In fact, some are designed for that very purpose, especially administrative charges or fees meant to pay the costs of a pollution control program. But their main purpose is to change behavior through a charge or tax that acts as an economic incentive or disincentive to alter conduct. (Although the words tax and charge are often used interchangeably, a tax is usually a levy meant to raise general government revenue, while a charge raises revenue to finance a given program. Water charges, for example, are usually used to pay for some aspect of water supply, such as purification.)

We feel that, as a general principle, the introduction of economic instruments should be guided by the doctrine of "fiscal neutrality": that is, the revenue does not increase overall government revenues. Revenue raised above and beyond funds needed to finance a given program can be recycled for other purposes, such as reducing taxes on such things as employment, investment, income, and savings. Where instruments prove regressive, affecting the poor disproportionately, then part of the revenues can be used to correct this effect.

Also, the introduction of economic instruments should be accompanied by a parallel package of measures designed to encourage the desired change in behavior. Pricing unleaded below leaded fuel is more effective, for example, if it is accompanied by policies encouraging car manufacturers to produce cars that run on unleaded, petroleum companies to offer unleaded at all gas stations, and environmental groups to educate motorists about converting old cars so they can use unleaded.

For an economic instrument to work, it must affect behavior. This means that it must move toward a threshold level of cost that will make polluters and consumers change their decisions about which processes to use and which products to buy.

A government requirement that information be made public comes under the heading of a market instrument because it enables consumers to choose, and it can be effective in altering corporate behavior. In the United States, industry was required to disclose the nature and amount of emissions, even legal ones. U.S. environmentalist Jeffrey Leonard believes that this requirement has done more than any other regulation to clean up industry.[3] It allows the public to see the records and act accordingly; it ranks companies, and it demonstrates waste to boards of directors.

Advantages of Economic Instruments

Environmental policy aims to protect the environment by altering and developing technology toward less polluting and less resource-intensive technology, by changing management behavior, by adjusting the structure of industrial output toward more efficiency and less pollution, and by changing consumer preferences. To achieve this, business wants an optimal mix of command and control, self-regulation, and economic instruments, and we believe that the present mix in most countries relies too heavily on the command approach. It is thus entirely appropriate that many governments are now studying and introducing economic instruments, which have two important advantages.

First, firms' compliance costs tend to be lower with economic instruments. The government's administrative cost may be lower as well. In air pollution, the cost of direct control is often from 2 to 20 times more expensive than the use of economic instruments.[4] This is because the command-and-control approach does not encourage companies that can clean up at low cost to clean up more than companies for which similar steps are very expensive.

Second, economic instruments encourage innovation. They encourage polluters to change to cleaner technologies and to develop new technologies because it pays for them to clean up more, in particular if pollution rights are tradable. They encourage new entrants to try to gain a competitive edge by starting off with new technology. Command approaches can have the same effect, but as they often require companies to use a specific technology, they may not be as effective in motivating continuous change and improvement. In fact, regulations based on outmoded technologies can actually have the effect of slowing improvement in an industrial sector.

Neither the command nor the economic instruments approach should affect competitiveness within a nation for reasons other than environmental efficiency. As compliance costs are usually lower with economic instruments, their use in one country may make an industry there more competitive than the same industry in a country relying on regulations to achieve the same effect.

The command approach often seems more convenient because both business and governments have much more experience with regulation, and relatively little with economic instruments. So both may feel comfortable dealing with commands. Governments tend to favor a command approach because of this tradition and because they feel that regulation is more certain. Business has favored regulation in the past because it also is more familiar with this approach, and feels it can influence it through negotiation. In addition, in many nations regulations are passed but rarely enforced.

Business also fears that instruments might simply be added on to regulations in place, and that it will thus face a double burden. This should not happen. If a regulation is working, there is no need for an economic instrument. But there may be cases, under an optimal mix approach, where a regulation is used to guarantee health and safety standards and an economic instrument encourages continuous improvement.

A 1991 public policy study sponsored by U.S. Senators Timothy Wirth and John Heinz concluded: "Over the past two years, we have witnessed dramatic changes in the political landscape of environmental policy.

Legislators, bureaucrats, environmentalists, business persons, and citizens of all kinds have come to recognize that market-based instruments belong in our portfolio of environmental and natural resource policies."[5]

That same report introduced the concept of economic instruments as follows: "Why all this emphasis on market forces, in the first place? The answer is purely practical. Selective and careful use of economic incentives can enable us to achieve greater levels of environmental protection at lower overall cost to society. A central principle is that as consumers and as producers, each and every one of us needs to weigh the full social costs and consequences of our decisions before acting....Market-based environmental policy mechanisms provide various ways to make consumers and producers recognize these social costs and consequences, and thus provide incentives for environmental protection. The creativity and power of the market—the awesome strength of millions of decentralized decision-makers—can be deployed on behalf of environmental protection, instead of against it."[6]

The argument most often used by environmentalists against economic instruments, especially tradable pollution permits, is that they are "licenses to pollute." This argument carries no weight, for any command-and-control regulation that allows any pollution at all can also be described as a license to pollute.

Fair and Equal?

The use of economic instruments raises questions about the distribution of their impacts. If economic instruments have regressive effects—placing greater hardship on the poor than on the rich—then compensatory measures may be required. If a particular industry or sector is harder hit than others (for example, those that are highly energy-intensive), then transition measures to allow for the necessary changes may be required. Yet ultimately we have to accept that a move toward sustainable development will cause far-reaching change in the structures of business and industry; there will be losers and there will be winners.

Most of the current environmental taxes and charges on products in OECD countries are relatively small, on such items as shopping bags and batteries, which do not constitute a high proportion of any family's income. Thus they do not raise issues of distribution of effects.

But taxes on products that meet basic needs do raise such issues. A study by Britain's Institute of Fiscal Studies (IFS), for example, found that the poorest British families spend a higher proportion of their income on energy than do wealthier families, so a charge on energy would hit them harder. Also, such a levy would provoke smaller reductions in energy use in richer households, which use more energy. But a tax could also produce returns that could be used to compensate groups that society wishes to protect.[7]

Since the idea is to change behavior, the best approach may be to charge the poor the same as the rich, but then redistribute money to the poor so they do not suffer unduly. They could receive credits or rebates. Many possibilities exist.

However, the IFS found that there were great differences between the efficiency of domestic energy use not only between but within classes. In other words, some poor households are more energy-efficient than other poor households, and the same is true of rich households. Thus the desire not to punish the poor must be balanced with the need to encourage changed behavior in all income ranges; large lump-sum compensations might decrease the incentive effect of a charge. Incentives can be provided by several methods, including schemes to use some of the charge to improve insulation in a house. Such schemes are best backed up by information and credit help. In the United States, some power companies offer to inspect every customer's home for energy efficiency, and to recommend improvements.

Experience So Far

A 1991 review of the prevalence of economic instruments in OECD countries found that their use is increasing rapidly.[8] Since a larger OECD survey in 1987, the number of charge systems either implemented or being contemplated increased almost threefold, and such systems were being used in 21 countries. There has also been a gradual shift away from using these instruments to raise revenues toward using them to change behavior.

Yet except for the Scandinavian countries, the use of such instruments—though growing—has remained limited. The OECD average was about six or seven instruments in use per country, with perhaps two or three water and waste user charges, a deposit refund system, and a tax differential on unleaded fuel being the standard spread.[9]

Economic instruments are used less in the developing world, probably because these countries have been struggling with issues of environmental management for a shorter time. Yet there is excellent work going on in the South that shows the scope for instruments there. At least three good reasons can be cited for developing countries feeling that they have much to gain from internalizing costs through economic instruments rather than regulations.

First, as noted, regulations are often more expensive. Southern environmental problems tend to involve resource use more than pollution, and these resource use problems tend to be spread across large rural areas rather than concentrated in cities, so the bureaucracy must be spread out as well. Thus economic instruments may be even more practical and cost-effective in the South than in the North. (Many developing countries, however, need to improve their institutional structure to make economic instruments work. And, needless to say, when large numbers of people do not participate in markets, as is the case with many of the rural poor, market instruments are not effective.)

Second, for developing countries reforming their tax structure, the potential revenue-raising side effects of most instruments may be helpful in decreasing direct and indirect taxes for the poor and taxes that discourage effort and investment.

Third, the South is in a position to study the North's long history of regulations and avoid the major errors. It is also in a position to monitor and copy—where appropriate—what works among the North's instruments.

But before new instruments can be put in place, two things should happen. Instruments already in place that actually encourage resource waste and environmental degradation should be removed. These include such things as tax breaks for clearing forests for farming, and subsidies on water, energy, pesticides, fertilizer, and so on. (Though in many developing countries, particularly in Africa, the problem is underuse, not overuse, of farm chemicals.) In many cases, the destructive subsidies are of most financial benefit to wealthier citizens who are able to use large amounts of subsidized water, energy, and chemicals, so removing them is "progressive." Any steps to do so, however, must be monitored carefully to see that the poor in developing countries are not made even worse off.

There are a number of basic preconditions for the effective operation of economic instruments. These vary from country to country but include such things as the establishment of firm property rights and

> *"Economic activity must account for the environmental costs of produc-tion. Environmental regulation has made a start here, albeit a small one. The market has not even begun to be mobilized to preserve the environ-ment; as a consequence, an increasing amount of the 'wealth' we create is in a sense stolen from our children."*
>
> William D. Ruckelshaus
> Chairman of the Board
> Chief Executive Officer
> Browning-Ferris Industries

contract enforcement, liability laws, policy reform, and institutional reform. The best instruments for the South will be those that place the burden of proof of compliance on business, as governments tend to lack the institutions to do the job. These instruments will also be designed to gain the confidence of industry by encouraging rather than punishing.

A scheme for hazardous waste management in Thailand, proposed by the Thailand Development Research Institute, shows how economic instruments can be used for a problem thought in the North to be best controlled by command and control. Hazardous waste producers are charged by an Environment Fund a fee that would cover the transport, treatment, and disposal of the waste, plus the same amount again as an incentive to reduce wastes. Half the total charge goes to deal with the waste; the other half goes into an escrow account earning interest. The Fund determines how much waste an industry is likely to produce based on records in other countries. Once all the contracted waste has been delivered for treatment, the money and interest in escrow are returned to the waste generator. The escrow fund thus acts as a performance bond to ensure delivery of waste. As firms produce less waste, they pay less to the fund. Performance is monitored by private auditing firms.[10]

Economic instruments have long been used in the management of living resources, mostly in the forms of simple fishing or hunting licenses, which both provide revenue for the resource managing author-ity and usually set a catch or kill limit.

Modern sensing and communications equipment allows for such management approaches to be much more sensitive and sophisticated. New Zealand, for instance, manages the fisheries within its 200-mile

Exclusive Economic Zone through the use of tradable permits to catch a certain quota of a given species of fish. A "paper chase," rather than inspectors, keeps track of catch amounts. Forms covering catches are filled in by skippers, permit owners, and processors, and then compared by computer. To base management more soundly on fish population dynamics, the quotas have been changed from tons of fish to percentage of allowed catch. Those fishing have generally welcomed the scheme— after hiring foreign scientists to check the accuracy of New Zealand government catch quotas.[11]

These examples of economic instruments within regulatory systems in the developing world and in the area of natural resources are only a small sample of the vast potential economic instruments have for protecting the environment.

The Right Mix

We need to ascertain where regulations work best and how they may be complemented by economic instruments and by self-regulation to form an optimal mix to achieve the objectives of sustainable development. The integration of command-and-control regulations and economic instruments is essential to avoid a two-tier, perhaps contradictory system. It would be highly inefficient, for example, to have one body imposing regulations and another setting charges in an uncoordinated manner.

We believe, and will be urging governments, that choices of mixtures of instruments, regulations, and self-regulation should be influenced by:

• *Efficiency:* The choices must be based on which measures work most effectively and most cost-effectively for society.

• *Flexibility of response:* Industry needs to be able to choose how to respond to regulation—that is, not whether to respond, but how to reach the target in the most efficient way.

• *Confidence in the regulatory environment:* Industry needs to know the nature and probable impact of regulation over significant periods of time so that it can plan its investments accordingly and not make technological investments that will be forfeited by rapidly changing regulations.

• *Gradual introduction:* Regulations should be introduced gradually so industry has time to plan its optimal response; if governments adopt the "precautionary principle," they will seek to introduce instruments quickly, but will give industry compliance time. If governments do not act until damage is advanced, there will be no time for gradual introductions. For its part, industry must accept that where urgent remedy is required, urgent regulations and enforcement are also necessary.

• A "level playing field":   Regulations should affect all comparable enterprises equally; while a level playing field can be established fairly easily under a single regulatory framework (such as national laws), countries at similar levels of economic development will need to harmonize their policies internationally to avoid trade distortions and unfair competition. The achievement of equity requires that different things be treated differently: a highly polluting industry will have to pay a higher pollution charge than a cleaner industry, but this inequality is the very purpose of the measure. Similarly, different levels of development between nations must lead to different standards and solutions in order to distribute burdens fairly and in proportion to the means available for corrective action.

• Transparency of compliance: It must be possible for each industry to be seen to comply with regulations; there must be no free riders or unduly privileged companies.

National Accounts

One way to support the internalization of environmental costs would be to alter the standard national accounts (SNAs) to reflect environmental damage and the depreciation of natural resources.

SNAs, upon which such indicators as gross national product (GNP) are based, were developed during World War II to allow governments to measure the effects of economic policies using standardized tools. GNP is a good instrument for many purposes and should be maintained, but it is faulty for measuring progress toward sustainable development. Not only does it offer no indication of environmental damage or the running down of natural resources, but the economic activity required to clean up pollution damage (such as oil spills) is paradoxically recorded as a contribution to economic welfare.

When a forest is felled for timber, GNP includes the income earned, but no loss of future productive capacity is recorded. If countries were run like businesses, there would be an accounting for depletion of valuable assets such as forests, oil, topsoil, and water. A proportion of income from the flow of a capital stock such as oil would be invested to replace income from oil as it became exhausted. In the developing world in particular, rapid depletion of natural resources adds to the GNP and suggests economic success, but no allowance is made for the reduction of income that will follow when the resource is exhausted.

The Norwegian government in 1974 and the French government in 1978 both established systems of natural resource accounting and budgeting. In 1973, the Japanese government introduced a new measure of net national welfare (NNW), which adjusts national income for environmental and other factors. Records show that while Japanese GNP increased by a factor of 8.3 during 1955–1985, NNW increased by a factor of only 5.8.[12]

Some economists argue that the best way to improve SNAs is to subtract from GNP not only capital depreciation but three other figures as well: money spent repairing and protecting against environmental damage ("defensive expenditures"), a monetary equivalent for any residual degradation remaining after defensive expenditures (such as soil erosion and acid damage to buildings not repaired), and an allowance for depletion of natural capital such as forests. This would produce a "net national product" (NNP).

It would obviously be difficult to derive figures for some of the categories above, especially residual degradation. It would also be hard to quantify in monetary terms irreparable environmental damage such as the extinction of species. So other economists argue that such an NNP exercise is not at present feasible, and certainly not cost-effective.

But SNAs can still be dramatically improved by the addition of known environmental damages that can be estimated in monetary terms and by incorporating net changes in stocks of natural resources. The effects of such changes in national accounts calculation could be dramatic. They could mean that when annual economic indicators were announced by governments, they would include an indication of how much given industrial, forestry, or mining operations cost the nation in terms of real natural assets. It could be more effective than any other measure in ensuring that environmental concerns were taken seriously by key ministries, such as those responsible for finance, industry, energy, mining, agriculture, and forestry.

U.S. economist Robert Repetto and his colleagues showed that when depreciation for oil, timber, and topsoil were included in calculations of Indonesia's national economic performance, economic growth between 1971 and 1984 was not 7 percent—the official figure—but only 4 percent.[13]

British economist David Pearce and Swedish economist Karl-Göran Mäler have predicted relatively rapid change in ways by which nations judge their performance: "The system of national accounts can be seen as a statistical system monitoring the conventional economic areas. The work under way in many international organisations—United Nations, UNEP and the World Bank—and in many countries will probably, in a couple of years, extend this monitoring system to include environmental resources, both in order to provide a better statistical data base for economic analysis but also to yield an improved wellbeing performance index for nations."[14]

Society's Choice

Recent history has reemphasized the advantages of the market system over other economic systems. The market finds the most efficient methods for creating wealth, and it offers the best chances for progress by unleashing human creativity. It offers society effective paths toward its goals; but society, through the political and legislative process, must set those goals.

So long as creating wealth ranked far above protecting nature as the major objective of society, markets guided economic actors such as consumers and producers toward maximizing wealth creation without concern for the resulting degradation of nature.

Today the world's nations—to different degrees and with different priorities—appear ready to base a reassessment of their long-term objectives on the realization that economic progress can only be achieved amid plentiful environmental resources and within healthy global ecosystems.

Changing course in accord with this new realization does not imply abandoning a system that has proved its merits. But it does mean that economic actors need the right signals to steer them toward the new objective. The nature of such signals should be as compatible as possible with the nature of the market system.

The fundamental signals guiding market decisions are prices reflecting the relative scarcity of goods based on supply and demand. With a given supply, increasing demand should result in rising prices. This basic mechanism has never been given a real chance to work for the environment. The use, exploitation, and degradation of nature has not created signals of scarcity because those who "own" nature and its services—society, expressing its wishes and intentions through government—have tended to give away environmental resources and services for free.

When severe problems and imminent threats finally required action, governments chose to guide behavior through command-and-control regulations rather than through price signals. This approach works against, rather than with, the market and weakens its main advantages: wealth creation, efficiency, and innovation. Thus business acts in its own interests in calling for prices increasingly to recognize and reflect ecological realities.

It will be a long and complex task to make prices reflect the ecological impact of resource use and the production of goods and services. This task is also laden with difficulties: assessing values and determining unknown costs associated with ecological impacts; changing basic elements of existing industrial structures; introducing potentially distorting elements in international trade, which may put a higher burden on the poor; and tempting governments to abuse a potentially important source of revenue, by using it for purposes other than guiding choices toward sustainable development.

In view of these problems, we appeal for a long-term orientation, for calculable and internationally harmonized policies, and for the gradual introduction of appropriate market signals.

Approaches not based on market solutions may present even greater hazards. The advantage of the market approach is that risks are offset by new opportunities inherent in competition, efficiency, and innovation. If these essential strengths of human ingenuity are not made to work toward the new objectives of society, we see little chance of development ever becoming sustainable.

3   Energy and the Marketplace

"We strongly advocate common efforts to limit emissions of carbon dioxide and other greenhouse gases, which threaten to induce climate change, endangering the environment and ultimately the economy."

Declaration of Group of Seven Leaders
Paris Summit, 1989

Energy offers some of the hardest challenges in the search for sustainable development. It is crucial for human progress. Its use gives rise to global, regional, and local pollution. And its price rarely reflects the environmental costs associated with its use.

This chapter is not an in-depth analysis of the energy question. Rather, it is an attempt by a number of business leaders to look at the issue from a business perspective in light of the principles discussed in chapter 2, seeking the right mix of economic instruments, command-and-control regulations, and self-regulation to improve the present blend of energy sources. Before going into the problems of internalizing external costs, however, it is worth looking at a few basic, inescapable facts that underpin the current debate.

Energy Dilemmas

The environmental impacts of energy production and use are complex. Sulfur dioxide and oxides of nitrogen contribute to local and transborder pollution in the form of acid precipitation. In developing countries, biomass (wood, charcoal, farm residues, and dung) provides 35 percent of all energy for cooking and heating for most of the population, but its use accelerates deforestation and the overuse of land.[1]

The threat of global warming is the most politically difficult issue in this field. About half of greenhouse gases caused by human activities are associated with energy, with the other half resulting mainly from defor-estation, the increase in grazing lands and rice paddies, and the use of chlorofluorocarbons and agricultural fertilizers.

We cannot return to the lower energy scenario of the past nor change our energy systems drastically. All countries have built their economies on an industrial infrastructure heavily dependent on fossil fuels, and a rapid change would have politically unacceptable economic impacts, especially in the emerging industrial societies of developing countries.

The largest per capita energy use is in the industrial world; the one quarter of the world who live here are responsible for two thirds of global energy consumption. But energy use in the developing countries will grow quickly because of rapid population growth, demographic changes, and economic development. Less than 25 years from now, total energy consumption in developing nations might pass that of industrial coun-tries.

A Framework for Action

Past energy policies have attacked the symptoms rather than the real causes. Society does not ask for energy, but for convenient services; it demands comfort, not heaters or air conditioners.

Given the complexity of the energy issue, we must describe not a final target but a process that allows for and fosters a continuous change in energy use patterns; a move toward more efficient production, transmis-sion, and conversion of energy into social goods; and continuous re-search and education.

Some "no-regrets" policies should be initiated without delay. These are actions, such as improving energy efficiency, that make economic and environmental sense whether or not global warming becomes a dangerous reality. But they should be combined with longer-term ac-tions coordinated on a global level. We need energy policies that are effective both locally and globally, that can adapt to new scientific understanding, that are both flexible and consistent, and that involve the public and private sectors—industry and consumers alike.

We propose therefore a reorientation of national energy plans toward a rational and coherent resource policy with a longer time horizon, built on three pillars, with a large supporting role for industry in each case.

(See appendix I for details of actions that will have effect over the short, medium, and long term.)

The first pillar is increased energy efficiency, which can bring the quickest returns and can buy time for longer-term actions. The members of the Organisation for Economic Co-operation and Development (OECD) have proved during recent years that economies can continue to develop while conserving energy, as their energy consumption per unit of gross domestic product has declined while their economies have grown. There is much room for substantial improvements in efficiency everywhere.

The second pillar of a rational energy strategy is a transition toward a more sustainable mix of energy sources and consumption patterns. This must happen in a careful and systematic way, to avoid major damage to economic development. A new mix of energy sources must also consider potential impacts on the local and global environment and on development demands in the developing world and Eastern Europe.

The third pillar is a long-term energy strategy for developing countries. This requires the development and use of indigenous resources, and reform of energy pricing policies. As industrial countries can help these nations get access to state-of-the-art equipment, technology cooperation is a key ingredient in a global energy strategy. (See chapter 8.)

Each country must determine its own mix of policy instruments, as discussed in chapter 2. Business will play a major role in implementing the new policies, which will only succeed if they are cost-effective, technically realistic, and consistent with agreed global objectives. We therefore urge governments to formulate a rational policy framework that encourages both industry and consumers toward sustainable energy development.

Making Energy Markets Work

Historically, energy policies have focussed on economic rather than environmental considerations. In some countries, energy use is highly subsidized, either directly or through tax breaks granted for a variety of reasons. On the other hand, in most countries taxes on fuels or energy are a major source of fiscal revenue. From now on, the environmental damage associated with energy must help define policy in this sector.

Basically, energy prices that do not reflect full costs have encouraged wastefulness while delaying progress toward greater energy efficiency and a cleaner mix of energy sources. A comparison of the energy use patterns in the former Soviet Union, the United States, and Japan shows that low energy prices make consumers generally less concerned about the efficiency of products. The time has come for energy prices to reflect full environmental and economic costs. This does not simply mean adding new taxes. Policy reforms should concentrate on more evenly spreading the burden of existing tax policy, removing subsidies, and internalizing in the prices of goods and services the costs of using environmental resources.

The environmental externalities of energy remain highly uncertain, but this is no excuse for inaction. The direction of change is more important than the quantity. A case can be made for a small initial carbon content charge—perhaps thought of as a kind of insurance premium— that could be increased (or reduced) as knowledge of the problem and its costs accumulate. The overall effects of such charges must be carefully considered before any government takes unilateral action that distorts the competitive playing field. International agreements on common environmental goals can help ensure that national policies are both cost-effective and environmentally effective.

Command-and-Control Regulations

A command-and-control approach has clear limitations in the energy sector. Regulating an increasingly complex industry requires a high level of technical competence, which often only industry itself possesses. Regulations can also be inflexible and expensive to administer, and are often inefficient in reaching environmental objectives.

Yet government regulations do not need to be specific and direct, but may be based on performance goals. Defining energy performance standards for future car engines in the United States, for example, the Corporate Average Fuel Efficiency (CAFE) standard was a major factor in doubling the average efficiency of cars made there between 1973 and 1987. Through CAFE, the government set a standard of performance and then left it to industry to find the best ways of reaching that level. Such an approach can be effective, as long as standards are set in close cooperation with industry.

Public-Private Partnerships and Self-Regulation

Self-regulation is becoming a more widely accepted instrument of change, especially as industry begins to take up the challenge of sustainable development. Industry associations can play a useful role in setting standards that trigger competitive forces. Such standards have already played an important part in the considerable energy savings of the aluminum industry.

Communication between enterprises, often promoted through trade associations, helps improve industry standards. How well enterprise or industry sectors do in reaching their goals can be confirmed by energy audits, with performance related to sector-determined standards or "best practice" rules. More formal International Organization for Standardization standards on energy use, or goals set by industry itself, may be an effective operational initiative to increase energy efficiencies in the short term.

Alternatively, the corporate adoption of guidelines that commit management to run the corporation in an environmentally sound way could serve the same purpose. Customers, often the most efficient influence on corporate behavior, should help corporations live up to their commitments by favoring energy-efficient alternatives.

Energy labelling of products and energy ratings of buildings and homes would help raise consumers' awareness of these issues. Governments can achieve this by adopting standards for mandatory disclosure of easy-to-understand information on energy consumption.

Economic Instruments

The right market signals can accelerate the move toward a more sustainable energy economy, as industry changes most efficiently when it is responding to market signals or to what it perceives as the future development of such signals.

At the moment, many government incentives and subsidies distort the market. Examples include direct price subsidies of energy, coal subsidies in Europe, electricity tax exemptions (such as the lack of a value added tax on electricity in some European countries), tax concessions for company cars, and undercharging by electrical utilities (which is widespread in developing countries).

> "In our industry, we have to carry on pursuing excellence in our environmental performance. We must continue to find and extract oil and gas, and make products, in ways that create less pollution and pose minimal threats to the global environment. We must also add our knowledge, expertise, and experience to the body of information that is being complied worldwide to combat global pollution."
>
> Lodewijk C. van Wachem
> Senior Managing Director
> The Royal Dutch/Shell Group

Thus the first priority must be to abolish subsidies so that the prices at least reflect the full economic costs of energy. In a transition period, incentives may play a useful and corrective role to ease the move. Soft loans or grants can encourage private energy consumers to invest in energy conservation, and special funds may help companies in developing countries to embark on efficiency programs.

Charges and taxes on energy usage and carbon emissions have received considerable attention in Europe and several industrial countries outside it as a way of curbing energy consumption and carbon dioxide (CO_2) buildup and of encouraging the development of new and renewable energy sources. International Energy Agency calculations show, however, that realistically high energy prices alone cannot achieve the goal set a few years ago at a semi-official meeting in Toronto of stabilizing CO_2 concentrations at 1990 levels, not even in industrial countries.[2]

Even if increased energy prices may not be effective in the short term, they have an important role as a signal over the long term. An expectation of increasing prices will influence people's habits, change corporate investment decisions in the medium term, and fuel technical innovation in the long term. Additional benefits arise if revenue is earmarked to specific environmental research and development.

Marketable emission permits, another example of economic instruments, are receiving serious consideration for selected national pollution problems, such as controlling acid rain or greenhouse gases. In practice, a total emission limit is set for an administrative region, and companies are allowed to trade emission permits, selling them as their emissions fall. Thus the market is used to encourage industries to determine their own most cost-effective ways of reaching a target.

This approach also has applications in technology cooperation deals. A North American coal-burning electric utility, for instance, might have to pay carbon emission charges or buy tradable permits, but would also be allowed to earn credits by lowering CO_2 buildup through planting trees in Brazil or by increasing the efficiency of coal-based power production in China. Used effectively, this approach could reduce emissions at the lowest cost while encouraging North-to-South investments, technology cooperation, and training.

Choosing a Balanced Mix of Instruments

The best energy policy mix will be one that allows both producers and users to benefit by choosing a sustainable energy path. The goal should be price-induced conservation by the consumer and cost-induced innovation by the producers of energy and energy-consuming products.

The most important criterion in selecting policy instruments must be their cost-effectiveness, or economic efficiency, in achieving the desired results. A mix of instruments, adopted to the specific conditions of each country, will maximize cost-effectiveness. Voluntary initiatives should be fully used and be coordinated with a rational use of economic instruments. Where effective, the mix should also be complemented with government regulations based on performance-based standards.

A gradual phase-in of these new policies is needed to ensure that information is provided well ahead of time, so that industry can prepare for the changes. To reach these objectives, governments must define the policy framework in close dialogue with industry. Some developing countries may need additional time to adjust their economies or additional resources to meet common goals.

Increased Energy Efficiency

Energy efficiency has been a stated goal of most governments for about two decades. But policies have not supported this goal. Once the right policy mix is put in place, astounding savings in energy and capital could be achieved at the same time that pollution is being decreased.

Increased energy efficiency has the largest potential for short-term contributions to sustainable development. This "no-regrets" policy is the major immediate option for reducing environmental costs without reducing the benefits derived from energy use. Industry's main contribution will come in realizing the full potential of this approach.

Studies by the Intergovernmental Panel on Climate Change (IPCC), established in 1988 by the U.N. Environment Programme and the World Meteorological Organization, are based on work done by almost 1,000 researchers and experts from some 60 countries. The IPCC and the International Energy Agency have assessed the reductions in energy use that could be achieved through increased energy efficiency and conservation in different sectors of society.[3]

The BCSD Energy Working Group has done a similar study, which has gone further and suggested actions for businesses as well as governments to remove barriers to efficiency. Our findings consider the potential of efficiency in electricity generation, the commercial and residential sectors, industry, and transportation.

Electricity Generation

Electricity demand is expected to grow rapidly in developing countries, in tandem with economic growth and industrialization. But electricity's share of all end-use energy will grow in highly industrialized nations as well. In the United States, electricity accounts for nearly 40 percent of the energy used, which is expected to increase to 60 percent by 2010.[4]

In the future, electricity could also penetrate the transport sector through electric vehicles or, alternatively, be used to generate hydrogen fuel. Increased demand for electricity means that new power plants will be needed, making it possible to realize considerable efficiency improvements. The IPCC estimates that efficiency improvements of 15–20 percent over average existing coal plants could be achieved through retrofits, and up to 65 percent through new generation equipment.[5]

The main reason for the often low efficiencies of existing power plants is the very long operating lifetimes of plant components, normally 35–40 years. The difference between today's average operating efficiency and the best available equipment thus reflects technical and commercial improvements during the last 15–20 years.

The potential for continued improvement remains substantial. The operating efficiencies of power plants in industrial countries should be at least 20 percent higher for plants ordered in 1995 than for those ordered in 1987. Major breakthroughs for completely new technologies, such as fuel cells, could accelerate efficiency gains. Efficiencies of 60–70 percent are conceivable, although not until well after the end of the decade.

Thus the main barriers to efficiency in electricity generation are availability of investment capital, long investment times, and price subsidies that discourage spending on efficiency. To deal with these barriers, which are similar around the world, we propose the accelerated introduction of state-of-the-art technology in production, conversion, and transmission; improved plant management; improved load management concepts; and efforts to give the right signals to consumers and to avoid distorted prices and subsidies.

OECD countries have generally come further in dealing with these problems, but large improvements can still come from better load management through least-cost utility planning, education and information, and more realistic electricity prices to finance better technologies. Policies promoting these elements should increase energy efficiency in the electricity sector worldwide.

In many newly industrializing economies, power plants are relatively modern and therefore efficient. The situation can be further improved in these countries through better plant operation and improved infrastructure. In most developing countries and in Eastern Europe, on the other hand, great contributions may come from an improved infrastructure for transmission and distribution. Losses in transmission are up to 30 percent in some developing countries, compared with 8 percent in the United States and 7 percent in Japan.[6]

Generally speaking, electricity prices reflecting the replacement cost of new plants—that is, marginal cost rather than average cost from older plants—would encourage improved efficiency on the demand side and help finance more-efficient power plants. It is on the demand side that energy efficiency improvements can be found at least cost, so major efforts should be directed toward developing energy standards and improving efficiency in end products.

Commercial and Residential

The IPCC estimates that new homes could be twice as energy-efficient as existing ones, and commercial buildings, 75 percent more efficient.

Retrofitting existing homes could lead to average improvements of 25 percent, while doing the same for commercial buildings could cut energy use in half.[7]

Several factors obstruct a rational energy policy in these sectors. For example, policies often do not reach the right decision makers. Builders, not home buyers, usually decide what appliances and heating systems to install. Not surprisingly, they focus more on initial than operating costs, and they choose less expensive, energy-inefficient equipment.

More-efficient products, such as better insulated houses and better heating, ventilation, and air conditioning systems, often do not find buyers due to higher purchase prices. Requirements to inform home buyers or future tenants of expected energy costs and potential savings may get around this problem. It may also make consumers more knowledgeable about operating costs, and they would choose to spend more for efficient equipment that would pay for itself through energy savings many times over.

We propose removing the barriers to efficiency in the commercial and residential sector through improved appliance standards and building codes; better energy information and declaration at the point of purchase; purchasing routines based on lowest life-cycle cost or lowest annual cost, instead of lowest first cost; grants, loans, and tax incentives for energy-saving installations; and individual consumption measurement systems in multifamily houses and other buildings with several tenants.

Educating and training professionals (particularly architects, designers, and builders) in how to construct more energy-efficient buildings would also be a major contribution.

Industry

Industry accounts for more than one third of energy consumed worldwide and uses more energy than any other end-user in industrial and newly industrializing economies. According to IPCC estimates, energy efficiency may be increased by 15 percent in some subsectors of industry and by more than 40 percent in others.[8]

Capital, labor, and energy all affect one another in industry. A price increase in one may encourage the substitution of more of the others. For energy-intensive industries—where energy expenditure is a large proportion of a company's total expenditure—increased efficiency has always been seen as a way of reducing costs.

In order to save energy and thereby reduce energy costs, industry has developed ways to recycle products and to use wastes as fuel. In the cement industry, a growing share of the total energy used is derived from waste oil. In the pulp and paper industry, wood residues are a major energy source. Secondary, or "used," aluminum accounts for 27 percent of every metric ton of aluminum consumed in the West. Recycling aluminum saves up to 95 percent of the energy needed in the initial reduction process.[9]

Cogeneration—the production of electricity and needed heat at the same time—is another way for industries to use energy more efficiently. However, national policies should be designed to support this approach. A good example is the U.S. Public Utility Reform and Policies Act of 1978, which made it possible for industries to use cogeneration more effectively.

Long equipment life creates investment horizons of 20–30 years, while more-efficient equipment becomes available yearly. Thus the lower the energy prices, the longer old and inefficient technology remains in use— particularly in energy-intensive industries.

Industrial energy consumption patterns suggest three stages in energy conservation: housekeeping (maximizing the efficiency of existing technology and processes), retrofit, and process technology improvements. In industrial countries, price-induced improvements in energy efficiency have been achieved largely in the first two stages.

Developing and centrally planned economies can improve efficiency at the very first stage. Investments in training workers in energy efficiency can pay for themselves quickly in savings of both energy and money. As capital equipment has a long life cycle in these regions, the technical efficiency of most plants is rather low. Many industries in developing countries may use more than twice as much energy to produce a unit of output as factories in industrial countries do; often this is encouraged by subsidized energy prices from national energy companies.

Transportation

Transportation is a major and rapidly growing energy user, responsible for about 20 percent of global fuel use, and in industrial countries for as much as a third of total energy consumption.[10] Present growth rates suggest that the world fleet of 500 million road vehicles could double by the year 2030, increasing energy and resource consumption and exacerbating transport-related problems of waste, noise, pollution, and excessive land use.

Major restrictions of transportation volumes is not a viable solution, since access to efficient transportation is vital for economic development. Limiting transportation in developing countries would impede the integration of these countries into the world economy. The goal therefore is improved efficiencies to minimize material and energy use. This will require both new technologies and a more intelligent use of those already available. Vehicle efficiency must be increased, and vehicles must be incorporated into a system that optimizes their specific advantages while reducing unnecessary or low-volume transport.

Industry's task is to create appropriate technologies and systems to cut resource use to a minimum. Providing the framework for action is government's responsibility. Government policies must include supportive market instruments encouraging environmental innovation.

The lifetime energy use of automobiles is best reduced through improved fuel efficiency. Manufacturers have already made considerable technical progress, offering new cars that consume only half the fuel of the present average car. But consumer demand for such vehicles has been low. Although the oil crises of the 1970s temporarily raised interest in fuel efficiency, people are once again demanding larger and more powerful cars.

Other barriers to sustainable transport include low efficiency standards on vehicles, low energy prices in some countries, lack of realistic alternatives to private cars, and tax breaks and subsidies for commuting in cars.

Stricter vehicle standards could cut energy use in the short term but would soon be offset by increasing numbers of vehicles. A wider variety of policy options, compatible with each nation's needs and cultures, is needed. Better traffic management offers both economic and environmental benefits. Smoother traffic flow can save as much energy as improving vehicle efficiencies does. New information technology can help through such innovations as "intelligent" traffic lights and computerized car controls. But improved traffic flow must be accompanied by other measures to encourage the use of public transport.

Cars and other vehicles can also be made more efficient through greater use of lighter materials and better thermal efficiencies. In the long term, vehicles must be developed that run on cleaner, more efficient fuels. Manufacturers are conducting substantial research on the use of methanol and electricity, but these new technologies are generally not yet competitive in terms of price and convenience.

Other modes of transport must be made available or more competitive. Rail transport, a low-energy mode that can be further improved, must be better used. To attract more industrial customers, service must be more reliable and punctual. International rail technology standards can make rail more competitive over long distances. Air traffic is the most sensible mode for long-distance passenger traffic; its environmental impact can be reduced through improved fuel efficiencies and more-efficient air traffic control. The capacity of truck fleets, well suited for distribution of goods, can be increased through use of computer-based information and communication systems. Sea transport, most efficient for long distance, intercontinental cargo exchange, can also be improved by information systems, leading to more joint use of containers worldwide. Europe's extensive canal system also needs to be better used.

Treating transportation as an integrated system will encourage better coordination between modes of transport. Such an approach would include providing airports with rail access, building park-and-ride facilities at railway stations, and creating central cargo terminals where goods are coordinated and bundled for different delivery systems. Again, such innovations are limited by costs, and calculations will favor them when environmental and social costs are included with monetary costs. Once this is done, planners can more rationally seek to minimize total transportation system costs. All of this requires new forms of cooperation within business and between business and government.

Some transportation can even be avoided completely through electronic communication in the forms of telecommuting and teleconferencing, but widespread use of such technologies will probably depend on increases in transportation costs.

The special transport problems of developing countries require urgent attention. In many, increased traffic is overwhelming underdeveloped infrastructure. Adapted know-how from industrial countries could help

reduce environmental impacts. This is a domain for technology cooperation joint ventures between corporations, regions, and governments.
Transport in developing countries could also be improved greatly through
a total systems approach, especially where private cars do not yet
dominate and there is greater scope for the creation of attractive mass
transit systems.

Local governments must give higher priority to transportation efficiency when planning land use and city expansion, and encourage
community and business participation in doing so. For example, better
integration of residential and business areas would cut commuting
distances. Traffic could be limited in commercial areas by simultaneously providing park-and- ride facilities outside the area and restricting car access to the city center.

National governments must give transportation the environmental
price tag it deserves. The sum of all taxes on cars rarely covers social and
environmental costs incurred by the road system and cars. But charges,
tolls, and marketable permits should be set with the entire transport
system in mind, and not be applied ad hoc to specific modes of transportation.

Governments should improve the competitiveness of public transport,
and ensure that different modes of transportation are allowed to compete
on equal terms and that sufficient funds are available for research on
energy efficiency. When investing in infrastructure, governments should
avoid past mistakes of favoring road and air transport.

A Sustainable Energy Mix

To achieve more sustainable energy use, efficiency gains must be complemented by a transition toward a more sustainable mix of energy sources
accompanied by changes in consumption patterns. This is the second
pillar of the rational energy strategy.

Fossil fuels are still expected to be the main source of energy well into
the next century. All fossil fuels—coal, oil, and gas—are associated with
carbon emissions. Coal will likely remain the foundation of global energy
use for some time: at 1988 rates of use, stores of coal, natural gas, and oil
could last 1,500, 120, and 60 years, respectively, although production
from unconventional sources could double the time for gas and oil.[11]

"If we are some day to have a truly sustainable world economy, we must begin to imagine a world chemical industry less dependent on petroleum. That could have broad implications for research, for the future of materials science, and for recycling and related technologies. In a truly sustainable economy, an energy company would seek to satisfy energy needs through a combination of renewable and nonrenewable resources and technologies. Energy development would mean more than seeking new sources and markets for traditional fuels."

Edgar S. Woolard
Chairman of the Board
E.I. du Pont de Nemours
and Company

Burning coal releases the most carbon of any fossil fuel. New technologies can increase efficiency and thus reduce emissions significantly, but many developing countries are unable to finance such technologies.

Oil releases less carbon than coal but more than gas. Supplies of oil and gas are less abundant, but adequate to meet expected demand well into the next century. Gas use is expected to increase more than oil, but neither of these fossil fuels can be expected to take over the energy share of coal.

Nuclear power currently provides 4 percent of the world's energy supply and 16 percent of its electricity.[12] The concern about global warming has brought nuclear alternatives back into the discussion and could make nuclear energy a promising energy source in countries with the infrastructure and know-how to operate such plants safely. New concepts for safer reactors and an acceptable solution to the waste disposal problem could encourage renewed public acceptance.

However, the risk of a severe accident is especially high in Eastern Europe, where many of the oldest nuclear plants have not only been poorly maintained during their entire lifetimes, but also have fundamentally unsafe and unstable design characteristics. This situation poses threats to people and property, and also to the future of safe nuclear power in general. It would be ironic if the use of nuclear technology were put at risk due to a second accident in Eastern Europe caused by technology not used anywhere else. The potential for increased energy efficiency in Eastern Europe, however, as well as the gradual change of the industrial structure in these countries, provides a great opportunity to shut down the unsafest units, with only partial replacements. This will require extensive technology and financial cooperation with more-advanced industrial countries.

Renewable energy sources provide 21 percent of world energy supply. Hydropower is responsible for 6 percent of the supply and 20 percent of the electrical generation. Only 20 percent of the worldwide economically viable hydro potential is used today, and in developing countries, only 10 percent.[13] Even though the development of hydropower can have its own negative impacts on the local environment, it offers a way to reduce greenhouse gas emissions significantly.

The cost of solar thermal, wind, and photovoltaic energy sources decreased significantly in the 1980s, with development stimulated by the oil crises in the 1970s. With sufficient funding, further large cost reductions are expected. Increased demand for renewable systems will also lead to economies of scale in production, making all, especially solar energy, more cost-competitive.

Organic materials like wood, crop residues, and animal waste can serve other than domestic purposes and be used as sources of heat, steam, or gas for electricity and be fermented to alcohol fuel, but today few if any large biofuel programs are economically viable. Developing countries could benefit greatly through better management of biomass resources. (See chapter 9.)

In industrial countries plagued by surplus food production, an alternative use of land for energy crops should be considered. Most of these renewable energy sources are widely dispersed, however, and thus large areas must be covered to generate sizable amounts of energy. This means that they are mainly a local contribution.

Waste is a source of energy that has, so far, not fully been taken into account. If the United States burned all its waste, it could meet about 5 percent of the country's total energy demand. So far only Japan, Switzerland, and Sweden burn more than 50 percent of their wastes.[14] Incineration may also reduce global warming, as it decreases the release of methane by landfills. Fear of uncontrolled emissions of carcinogenic substances, such as dioxins, has contributed to low public acceptance of waste incineration, but new technologies decrease such emissions. To the extent that recycling of wastes is economically and environmentally more attractive than incineration, recycling is naturally to be preferred.

More international trade in electric energy would reduce the need for peak generating capacity as well as reserve capacity in each region. We hope that governments will begin legal reforms that make such trade legal, thereby optimizing overall patterns of energy generation among countries. Such cooperation between the United States and Canada is an encouraging example, as are similar efforts to take full advantage of the variety of energy sources in the Scandinavian countries.

In conclusion, goals for improving the present energy mix should address varying time periods. In the short term, the focus must be on reducing the environmental impacts of fossil fuels and on encouraging energy efficiency. Lowering the risk of operating existing nuclear power plants in Central and Eastern Europe will ease the immediate pressure in these countries. This is a matter of urgency.

In the medium term, we should benefit from the ongoing development of clean coal technologies, a balanced nuclear expansion, some form of biomass-based energy (if it can become economical), solar energy sources, and further hydropower development. In the long term, industry must cooperate in intensive research efforts that may make fuel cells a widely used energy conversion technology.

Given these efforts, we foresee a gradual shift to a more sustainable mix of energy sources. How smooth and how rapid this shift will be depends on the choice of policy instruments.

Energy Strategies for the Developing World

Energy use in developing countries varies. The poorest have not yet formed a basic infrastructure, and may move through a period of developing heavy, energy-intensive industries. The newly industrialized economies have already been through this phase, and are beginning to shift to lighter, less energy-intensive industries.

Energy mixes also differ, depending on national energy source endowments, international trade partners, and government policies. In Brazil, 90 percent of energy used in power generation is hydropower; in India, 75 percent is based on coal; in Malaysia, 55 percent is based on oil. However, biomass provides on average 35 percent of commercial and noncommercial energy in the developing world and is thus the main energy source for a large share of the population. In Bangladesh, biomass represents 70 percent of total energy consumption; in Ethiopia, it is as much as 94 percent.[15]

Differing needs within each country also require segmented strategies. Many rural areas, for example, may benefit from decentralized energy generation, including the use of nonconventional energy sources. Agricultural regions may depend on efficient distribution and usage of electricity for irrigation and small businesses. Industrial and urban sectors have other energy supply and demand issues, and hence options for clean and efficient technologies that are comparable to those in industrial countries. These large differences suggest the following priorities in the energy policies of developing countries.

• *Encourage development of indigenous resources.* Strategies are needed to improve the efficiency of biomass use. Many existing modes of biomass use are highly polluting, dangerous to health (such as wood smoke in small, enclosed homes), and contribute to deforestation and desertification. Technologies to improve wood-stove efficiencies and better management of biomass sources are available, and their use should be encouraged. Training and research could further improve the situation.

For developing countries, large-scale commercialization of charcoal and fuels from biomass plantations may be important. The National Oil Company of Zimbabwe, for instance, proved that biomass can be used successfully as a substitute for nonrenewable fossil fuels. It also showed that technology cooperation can be facilitated by systematically training the local work force, adapting existing technology to local conditions, and cooperating with local vendors.[16]

• *Reform energy pricing policies and remove subsidies.* Commercial energy is often highly subsidized in developing countries. Electricity rates are often less than half those in industrial countries. These artificially low prices discourage efficiency and conservation. Furthermore, subsidies consume scarce capital, making less money available to improve the efficiency of power plants, reduce transmission losses, and address local pollution problems.

Although these problems exist in all nations, they are more severe in developing countries. In India, the practice of providing unmetered electricity to farmers leads to gross misuse (such as overirrigation) and provides no efficiency incentives. Indian economists argue that if cheap power really does benefit the poorer farmers, a direct subsidy to these individuals would be better.[17]

Full-cost pricing would improve industrial efficiency in developing countries. Given the current political and economic situation, full-cost pricing of energy is difficult to implement, and is likely to cause major social and political problems. In spite of this, it should be the firm priority in any developing-world energy strategy to remove energy-related subsidies.

• *Cooperate on energy technologies.* Developing countries have the advantage of being able to see and—we hope—avoid the mistakes made by more industrialized countries. They may therefore leapfrog into a new generation of sustainable energy. World industry should recognize the unique business opportunities in cooperating with developing countries to define more benign energy development.

• *Develop energy strategies locally to meet local needs.* This is both efficient and builds local confidence. For example, in Kenya more than a half-million urban households are now saving money and time using the Kenya Ceramic Jiko, an improved charcoal stove that uses 50 percent less fuel than the traditional charcoal stove. The success of this model is mainly due to the involvement of the existing informal small-scale production and marketing sector; the participation of the end-users in design, developments, and field-testing; and the involvement of local researchers and grassroots environmental groups during the whole process. Similar programs have been implemented in Burundi, Madagascar, Malawi, Rwanda, the Sudan, Tanzania, and Uganda.[18]

Energy efficiency can also become an important strategic management vision within local corporations. For more than 20 years the Tata Group in India has been involved in energy efficiency improvements in all the group's companies, and has developed specialized teams of experts and economists who regularly review energy technology and its operational implications in their factories. They thus serve both as "gatekeepers" for new expertise from abroad and as a formalized channel of exchange for indigenous experience. Their efforts are further supported by Tata Energy Research Institute, which works on energy issues on a national and international level.[19]

Ways Forward

Energy efficiency in production processes has increased steadily in all industries. Yet the potential for large energy savings remains.

Even larger cost-effective savings can be made in new products.

Industry can best contribute to such savings through production of improved products such as cars, light bulbs, refrigerators, and so on. But a demand must be created for such products. With continued low energy prices, consumers are generally indifferent to the efficiency of the products they purchase, even though efficient products usually save a great deal of money over their lifetimes. A system for energy accounting needs to be developed in order to aid in the evaluation of the total "cradle-to-grave" energy aspects of materials and products.

A better mix of energy prices, stricter standards, and better information will help save energy worldwide. A correct mix of such changes will be the first step toward sustainable energy development—and would provide substantial economic benefit at the same time.

In industrial countries, efficiency is constantly improved to pursue competitive advantages. Higher energy prices will speed up this development. Subsidies distort price signals, which should be encouraging efficiency, and should therefore be removed. Greater energy efficiency can be promoted by a wider awareness in industry of the potential and of its economic benefits. Business leadership and business organizations can accelerate this process.

In developing countries, the barriers to improved energy efficiency include limited resources, high capital requirements, lack of modern technologies, operational problems, and distorted energy pricing signals. The main barriers to achieving efficiency are likely to be management skills, coordination, and communication. The poorer countries should strive to understand and avoid the mistakes that industrial countries have already made. Industrial countries can help them obtain the most efficient technologies and develop indigenous competence. Technology cooperation is therefore a key ingredient in a global energy strategy. (See chapter 8.)

Economic growth, industrial investment, and structural changes are necessary to move along this path. The role of local enterprise and of a functioning industrial sector is thus important, which in turn requires a policy framework that supports the entrepreneurial development of the economy in these countries, and that attracts investment capital.

Capital Markets: Financing Sustainable Development

"Like it or not, the days when portfolio decisions could be made in a complete moral and social vacuum are numbered."

Financial Times editorial, April 1990

Capital markets will play an important role in the search for sustainable development, but little is known about the constraints, the possibilities, and the interrelationships between capital markets, the environment, and the needs of future generations.

Virtually no research has been done on this subject. It is essentially new territory, a fact that forces us to limit ourselves to some very basic arguments and conclusions. We thus offer an agenda for further research, analysis, and dialogue.

Sustainable development requires increased investments in both industrial and developing countries. It requires long-term investments. And it requires more capital investments that respect environmental criteria and development needs in projects and enterprises. Most of the finance for such investments will have to come from capital markets.

The structure and function of capital markets affect the availability of capital, influence investment processes, and, by extension, influence the ways that business managers who are approaching investors project the current performance and future potential of their enterprises. Capital markets may be one of the most important factors conditioning corporate behavior, and are therefore a key issue in the future of the planet.

We have noted that sustainable development will require some of the greatest changes in industrial nations, where both production and consumption are highest, where the most pollution is released, and where capacity for change is greatest. As these nations begin to internalize environmental costs, and as the Polluter Pays Principle becomes more honored in practice, the ways in which their capital markets value corporations will begin to change. Companies that further the cause of sustainable development will be perceived as more valuable in the marketplace.

The emerging capital markets in the developing world are becoming powerful tools for development, and starting them off on a sound footing—making them more open, efficient, and competitive—should be a high priority for business and governments. Given the economic importance of natural resources, farming, and forestry in these countries, the investment and business community must grasp the importance of sustainability and see that it is reflected in the growing capital markets. Adequate information and market transparency are particularly important in this regard. The relative immaturity of these markets should not provide an excuse for environmentally irresponsible behavior.

Capital Markets: A Rough Sketch

First and foremost, capital markets transfer funds from savers to investors in productive assets, such as plant and machinery, and to providers of services. This separates ownership from management of these assets in the economy, ensuring an allocation of resources to those with a comparative advantage in using and managing them. Capital markets also provide a mutually beneficial bridge between individuals, who have short time horizons, and companies and projects, which have longer timelines. These links ensure that the investment patterns of savers support the long-term horizons required for capital investment.

Equity capital is provided to corporations through the mechanism of the stock market; short-, medium-, and long-term debt (essentially at fixed rates of interest, sometimes enhanced with equity features) is made available through the bond market; and short- or medium-term debt (usually with a floating interest rate) is provided through the banking market.

The characteristics of these three markets are largely determined by different investors' perceptions of the quality of earnings they will bring. Thus equity market investments are generally regarded as more volatile than bond markets or bank deposits, and it is their incremental yield over lower-risk alternatives that draws investors into stocks. Yields on stocks derive either from upward share price movements (capital gain) or from dividends. An investment decision favoring stocks is usually supported by detailed analysis of certain aspects of the company's business.

Bond markets, on the other hand, rely to a greater extent on investor perceptions of credit-worthiness and future movements in interest rates, with revenue accruing either from capital gain on the value of the bond or from the income stream of coupon payments.

Finally, bank deposits are usually related to depositors' shorter-term perceptions of interest rate variations, but also to longer-term commitments by the bank lending to a project or corporate borrower. Interest rates are related to the general availability of funds in the international or domestic financial markets. Depositor security depends to a great extent on the bank's lending policies, which, with its capital, underpin the solvency of the institution.

The intermediaries in all these transactions form the financial services sector. The constituents of this sector and their relative importance vary from country to country, but generally include banks, stock exchanges, brokers, and insurance companies.

Clear and closely refereed rules for capital markets are necessary and acceptable in market economies because all participants recognize that this ensures a level playing field for open competition, equality of information, and thus the confidence required for the markets to work well.

World Capital Market Imbalances

Capital markets are at different levels of development in different countries. The world's capital markets essentially work out of New York, Tokyo, and London. Related services, such as insurance or commodities exchanges, also tend to be centered around these markets. Despite vastly improved communications and technology to move market information, it is perhaps inevitable that major markets and services favor a few geographical centers, to the detriment of the interests of the developing world.

At the end of 1990, 95 percent of world stock market capitalization—totalling $9.5 trillion—was located in the United States, Japan, the United Kingdom, and the remaining major European stock exchanges, according to the International Finance Corporation (IFC). Of the remaining 5 percent ($470 billion), the combined capitalization of the emerging stock markets of South and East Asia amounted to $341 billion; that of Latin America, $78 billion; in Africa and the Middle East, $8 billion; and in other markets, $43 billion.[1]

Total annual stock market turnover in dollar terms for the same period saw the industrial markets accounting for 85 percent (of which 56 percent was in the United States and Japan alone), with Latin America representing a mere 0.37 percent and Africa, 0.15 percent. In comparison, the volume of publicly issued bonds in the world's major markets totalled $12 trillion at the end of 1990, with 64 percent for borrowers in the public sector and 36 percent in the private sector.[2]

One key reason for the slow start of domestic capital markets in the South and the reliance on borrowing was the availability of large amounts of bank lending, much of it derived from the revenues of the oil-producing countries during the 1970s. Also, many people in developing countries chose to invest their wealth in the industrial world—a phenomenon known as "flight capital." The emerging markets, although very small, have grown fourfold since 1985, and in 1990 provided $22 billion in new money for companies.[3]

It is now widely accepted by developing-country governments that institutional reforms of capital markets help make the investment climate more attractive. Some 40 percent of the external capital requirements of Indian private industry, for example, is now met through the Indian stock exchanges, in which 15 million Indians hold investments.[4]

This reflects newfound confidence among outside investors that a number of these countries are finally coming to grips with political and economic problems that have beset them since the debt crisis began. More important, it reflects the newfound confidence of their citizens. The flow of flight capital appears to have been stemmed and even reversed, much of this capital attracted back by the trend toward privatization. Once a source of embarrassment to some countries, these "hidden reserves" could rapidly become one of the principal motors of new growth.

The International Finance Corporation, the private-sector investment and lending arm of the World Bank, has accelerated this process by trying to reinforce the regulatory structure and depth of these emerging markets, to help promote interest in them in major financial centers.

By mid-1991, over 100 specialized investment funds had invested in these emerging markets, and foreigners held stock in them worth an estimated $17 billion.[5] In light of the financial power wielded by other institutional investors, these funds may prove to be merely the vanguard of major capital flows toward the developing world, driven by the quest for portfolio risk diversification. Making fund managers aware of the demands of sustainable development should transmit the message more effectively to these minor capital markets.

A whole range of basic political and institutional conditions attract or discourage local and international investments. One effect of the recent integration of the world's capital markets has been that Southern companies competing for capital are providing more information than before. Companies in the industrial world routinely provide large amounts of financial information, and are beginning to provide more data on environmental impacts, a practice that will eventually be taken up in the South. Likewise, new sensitivity to sustainable development issues on the parts of Northern institutional investors, banks, and insurance companies will also move throughout the integrated global markets.

Increased political stability is as important for attracting domestic and foreign savings as deregulation and the development of a reliable market infrastructure. In most cases, they tend to go together. (See chapter 10.)

Misconceptions and Constraints

Several misconceptions affect the relationships between capital markets and any progress toward sustainable development. The first is that the two subjects are unrelated. This misunderstanding has arisen largely because the public normally associates such concerns with issues of conservation—rain forests, endangered species, and so on. It has also been linked to pollution, so it is seen as a job for industry. But in the words of the World Commission on Environment and Development, sustainable development is not a fixed state but "a process of change in which the exploitation of resources, the direction of investments, the orientation of technological development, and institutional change are made consistent with future as well as present needs."[6] Capital markets play a role in all these processes, particularly the direction of investments and the exploitation of resources.

A second misconception is that sustainable development, concerned as it is with future needs, is only about long-term investments operating over decades—that it sacrifices short-term needs. This notion overlooks the fact that investment decisions are taken daily by thousands of corporations, funds, and individuals worldwide. The direction of those decisions is more important than the term of the individual investment.

> *"Business is the most forceful agent of change. As business leaders, we are responsible for change that is positive. This means to offer returns to owners, jobs and development opportunities for employees, quality services and products to suppliers and customers. But it also means to bring positive change to society at large over the longer term. This change must be a commitment to sustainable development."*
>
> Antonia Ax:son Johnson
> Chairman
> Axel Johnson AB

A serious constraint on capital markets' promotion of sustainable development is the lack of adequate information with which to evaluate companies and investments. This constraint, which permeates all capital markets—industrial and emerging—and all investor decisions, is discussed further in the following section.

Another problem is that worldwide demands for "sustainability investments" coincide with large debt burdens in the developing world, major needs for investment in Eastern Europe, and low savings rates in some major economies. The combined debt of Latin America, Asia, and Africa at the end of 1990 was $1.365 trillion, a figure roughly the same for four straight years.[7]

Constraints caused by capital scarcity and by the inability of many countries to absorb financing are even more complex in the developing world. Most such countries lack effectively functioning capital markets. Reasons vary, but include the combination of political and economic instability; poor governments and poor people; historical exploitation of natural resources by foreign corporations with little benefit to the host economy; inequitable distribution of wealth, resources, and political power; and, in many cases, ideological and religious opposition.

For a time, local corporations turned instead to international bank lending, which met most of their needs. The debt crisis of the early 1980s put an end even to this, leaving developing countries with an almost total withdrawal of financial support, steadily declining commodity prices, international currency and interest rate fluctuations, internal recession, and social upheaval. Such conditions—devoid of confidence and stability, the prerequisites of all successful financial markets—further postponed the birth and growth of national capital markets in which competitive access to capital could be guaranteed to a broad spectrum of borrowers and entrepreneurs. (See also chapter 10.)

Many existing private banks in the developing world are linked to private investor groups, and such links create private capital markets closed to all but a select few. Thus few small or medium-sized businesses have access to formal credit; they must rely on informal capital markets, with their associated high interest costs. In many countries bank lending has been so conservative (to finance cash flow or inventories rather than investment) and inefficient that it impedes the development of private enterprise.

Large debts and fierce competition for capital imply that environmentally sustainable development is likely to enjoy far lower priority in world capital markets than it does in public opinion and governmental programs. Only recently did multilateral institutions such as the World Bank realize that by directing their funds primarily toward the public sectors they have actually been discouraging the development of domestic capital markets. Recent IFC experience has shown that there is another way, and that the liberalization and reinforcement of capital markets in the developing world may be one of the most effective ways of improving economies.

Other impediments to reflecting sustainable development issues in capital markets are rooted more deeply in economic systems, and in the fact that since no one owns natural resources such as the atmosphere or the oceans, it can seem to be in no one individual's financial interest to invest in improving those systems. Internalizing the costs of degrading environmental resources will help to correct this.

Components of Reform

Capital markets comprise a complex system with an elaborate array of actors operating with an intricate and tangled set of motivations and goals. Improvements, therefore, can best be achieved by changing the perceptions of key actors in the markets, which requires that they have access to a great deal more information of much higher quality. The aims of sustainable development will only be served in the capital and banking markets by a significant change in the way the environmental performance of corporations worldwide is assessed for investment or lending purposes. The process requires a change in the relationship between borrowers on the one hand and investors and lenders on the other.

By and large, the market value of company stock is based not on what a company is contributing to sustainable development over the long term, but on its short-term earnings and dividend prospects, its corporate balance sheet, and its profit-and-loss statement. Hence to be successful in raising capital, management may be tempted to concentrate on short-term profitability rather than on securing substance and long-term potential.

But this generalization is not entirely true, and it also insults the sophistication of investors. The stock prices of many banks rose when they made provisions against debts in developing countries, even though this hurt short-term profitability. Investors are also aware that certain industries must spend heavily on research and development (R&D), to the detriment of annual results but in the interests of long-term competitiveness and profitability. In exactly the same way, investors will also have to understand that financing sustainable development increases long-term competitiveness and profitability.

Current methods of valuing companies are based on existing operations and future potential from new products and new markets. These considerations fail to account for, or at best undervalue, environmental costs and benefits in the long term. For example, the net value method discounts projected investment returns 20–50 years from now so heavily that they are virtually valueless. This is especially true in developing countries, where the uncertainty of the investment climate pushes the discount rate up. This often results in rapid consumption of resources rather than investment aimed at increasing the overall resource/result ratio. Scarce capital can be attracted to enterprises or projects that undermine rather than promote sustainable development.

Calculations of the internal rate of return in any given project are likely to prove more accurate over time if environmental costs are factored in at the project analysis stage rather than after unexpected cleanup bills have been paid. Such an approach entails new responsibilities for management but is more likely to reassure investors, even if it may lead to less spectacular economic results in the short term.

As companies pay more for their pollution and as laws change to increase the liability for environmental damage, a failure to assess fully

> *"We aim to generate appropriate financial results through sustainable growth and constant renewal of a balanced business structure, so that we justify the confidence of all those who rely on our company—stockholders, employees, business partners, and the public."*
>
> Alex Krauer
> Präsident des Verwaltungsrates
> Ciba-Geigy AG

the environmental bill for certain activities may lead to overestimation of a company's true worth, which could later translate into the collapse of the share price. From a risk management point of view, internalizing environmental costs would produce more accurate signals for individual investors, bankers, institutional investors, and insurance companies. It is therefore essential that the criteria and numerical values of sustainable development become an integral part of the information process, so that sound investment and lending decisions can be made.

Providing accurate numbers to reflect the reality of sustainable development issues depends on the type of accounting treatment given to such items as asset valuations and depreciation, future income streams, contingent environmental liabilities, and so on. As these may involve less tangible economic phenomena, accounting practice may be able to borrow from the valuation treatment given to such items as goodwill or patents, which have received serious examination by the accounting profession in recent years.

Such changes will require in-depth analysis and consultation with all interested parties: corporations, auditors, government tax authorities, financial analysts, investment managers, and bankers. Particularly important are amortization rules and the relative value given, for example, to resource-intensive as opposed to intellectually intensive activities. The current search by the accounting profession for new valuation methods can be of significant help in this effort.

Economists around the world are already working to address valuation issues, and government support for their work is increasing. The U.S. Congress has asked the Department of Commerce to develop methods for incorporating environmental quality into national accounts, and similar exercises are being conducted in some European countries. (See chapter 2.)

Company analysts have a critical role to play in the revaluation of corporate performance along sustainable development lines. Their adoption of environmental assessment techniques could be an effective way of bringing the positive elements of competition—innovation and action—into the environmental arena. And their direct involvement could be a swift route to ensuring that the market is able to assist the transition to sustainability. Just as industry is investing considerable capital in R&D, financial institutions may likewise invest in the development of more know-how, information, and improved methods of accounting and evaluation.

Investment decisions are based not only on the latest information and in-depth analysis, but also on certain simplified decision-making tools. The latter include rating services such as Moody's or Standard & Poors (S&P), which continuously analyze the activities of public and private borrowers, awarding them credit ratings expressed in terms of a points system ranging from triple A (best) to single C (worst). In fact, the statutes of many of the largest institutional investment funds specifically lay down the minimum rating quality of corporations in which investments can be made.

Eventually a similar system could be provided to investors and bank lenders through an analogous institution that continuously gives the "eco-efficiency rating" of a corporation after analysis of its activities and projects. This could lead to "triple EEE" ratings (environment, efficiency, and enterprise), serving to guide investors to favor highly rated corporations over less environmentally conscious rivals. The U.S. market will soon have the benefit of an environmental S&P 500 rating system, developed by the Investor Responsibility Research Center.[8]

Banks

Banks are only now beginning to appreciate fully the new opportunities presented by the sustainability challenge and the environmental liabilities lurking under the surface of the industrial activities they finance. Given the virtual absence of relevant information from borrowers, this is not surprising.

Unlike war or flood, however, environmental damage is rarely a case of "force majeure"; it can be foreseen. Today, there can be little excuse for failing to do so. In the United States, some banks have recently been held financially liable for environmental damage (usually unsafe disposal of hazardous chemicals) caused by companies they held following foreclosures.[9] For reasons of credit soundness, customer and investor relations, and liability, banks can no longer afford to ignore the environmental implications of their loans. An environmental audit is likely to eventually be an integral part of a loan application.

Banks can also play an important role as long-term shareholders in the industrial sector. This is the case in Germany as well as in the Japanese *keiretsu* system of interlocking equity participation. If this role were combined with information on the environmental effects of corporate action, such long-term perceptions of corporate benefit might favor the cause of sustainable development. However, the German and Japanese models currently provide less information to the public than other systems do.

Institutional and Individual Investors

Of growing importance among financial intermediaries in today's markets are the institutional investors, such as insurance companies, pension funds, and portfolio management specialists, acting on a fiduciary basis for many individual clients. Their significance lies in the sheer volume of funds under their management and the power this provides. For example, the U.S. pension fund industry currently manages assets worth more than $2 trillion.[10]

Institutional investors must become increasingly aware of the environmental risks and opportunities embedded in their investment decisions. They have the legal fiduciary responsibility to exercise prudence and due diligence. In many cases, particularly in the United States, this responsibility is being exercised as such investors take a more active role in company supervision, gradually moving from relative passivity to active influence on management and corporate strategy. If the message of sustainable development can be transmitted convincingly to these institutions, the cumulative effect could be enormous.

Corporate management, for its part, now pays special attention to these institutions, particularly in the provision of regular information about corporate strategy and developments. Given the right signals, the institutions holding pension funds may find themselves attracted naturally to the concept of sustainable development, as a long-term view on future values and an emphasis on capital appreciation corresponds to their basic nature and mission.

Even a small shift by these funds in favor of emerging markets could have a major impact on local growth in the developing world. Although clearly a new departure for many pension funds, such a shift might be prompted by the perception that the emerging markets may well turn in far higher growth and returns than established markets in the 1990s.

There is also a growing awareness among individual investors of the need to take account of environmental considerations. This concern is currently focussed mainly on more emotive issues, such as deforestation, the greenhouse effect, or endangered species, but it may in time become more sophisticated. Just as public opinion has been seen to oppose investment in certain countries for political reasons (apartheid in South Africa, for example), so it is probable that individual investors will increasingly take account of the environmental performance of individual corporations or economic groups in their portfolio strategies.

Insurance Companies

Insurers are in the business of assessing risk, and many are also large investors in capital markets. Environmental risks are becoming central to insurance policy valuation and the pricing structures of insurance premiums. Underwriters today confront enormous claims based on policies written decades ago, before environmental risks were recognized and factored into premium calculations. Companies with a dubious environmental record will find it increasingly difficult and expensive to get insured. Environmental audits, as with the banking industry, are likely to become part of the assessment tools of the insurance industry.

Environmental liabilities are also becoming one of the main concerns regarding investment in developing and East European countries. Insurance companies ask for relatively high premiums and in-depth risk assessments for such businesses. Methods for assessing risks and auditing are therefore being developed very fast.

Unsustainable development patterns appear to be part of the reason for the higher number of disasters over the past decades and the associated damage, injuries, and fatalities. First, particularly in the developing world, growing numbers of poor people are living in more vulnerable environments: areas prone to flooding, earthquakes, and landslides. Second, deforestation and overcultivation can increase the severity of floods, the disaster increasing fastest in terms of numbers of events, and of droughts, the disaster that affects the highest number of people yearly.[11]

There is also growing concern that climate change might produce more disasters—more storms and cyclones as climate systems are disrupted further by rising carbon dioxide concentrations in the atmosphere, and more floods as sea levels rise. This increased concern over the insurance risks inherent in unsustainable development should make insurance companies more sensitive to the needs of investing in sustainability.

Multilateral Lending Agencies

Fortunately, the lending policies of multilateral agencies such as the International Monetary Fund, the World Bank, the regional development banks, and their specialist corporate arms are increasingly taking account of the environmental dimension.

The World Bank alone increased its lending for environmental purposes fourfold between June 1990 and June 1991. Full or partial assessments of environmental impact were done for more than half the projects supported in 1990–1991.[12] The role of such agencies as catalysts for investment is particularly important, since through such means as cofinancing arrangements they can draw private-sector lenders and investors into the virtuous circle of sustainable project economics. In this sense, multilateral institutions have a leading role to play in providing more finance for sustainable development.

Governments

Official development assistance from the members of the Organisation for Economic Co-operation and Development to developing countries and the multilateral agencies totalled $62.6 billion in 1990.[13] As a percentage of the gross national product of recipient countries, the amounts varied from negligible to substantial. Thus a donor agency's potential to apply pressure in support of sustainability changes in direct proportion to the importance of the aid granted.

The attitude of governments will be critical to the development of capital markets that effectively support sustainable development. They must in particular address such questions as corporate and investment taxation, information disclosure, eligibility for export credit or regional lending, and all other factors directly enhancing the profitability or funding of corporations pursuing desired investment strategies.

The 1980s were marked by a shift of economic responsibilities from the public to the private sector: privatization has been the name of the game from London to Santiago de Chile. These changes will diminish investor perception of the state as project sponsor and borrower. They will also increasingly shift the burden of husbanding scarce resources and pursuing sustainable development strategies from governments to corporations—which will challenge them to contemplate longer investment horizons.

The state has many tools for directly or indirectly encouraging corporations, such as corporate and income taxes, depreciation allowances, and investment grants in support of capital expenditure fulfilling clear environmental ambitions. Its role as one of the managers of interest rate levels is also crucial. At the other end of the spectrum, governments have the power through regulatory or fiscal sanctions to punish corporations that neglect environmental and sustainable development policies.

Taxation is a particularly effective mechanism for sending economic messages to the business and investment communities. (See also chapter 2.) The subject is both vast and complex, and tax specialists will have to figure out how best to serve the overall goal of long-term sustainable development. Clear rules are necessary to facilitate the appreciation of capital, the building of savings and equity, and the possibility of smaller discount rates over a given time period.

Substantial change in legal and regulatory frameworks surrounding capital markets will call for international harmonization. Countries with developed capital markets have to cooperate with developing countries because a deep institutional reform of all capital markets is needed. Among the issues to be addressed are opening and deregulating capital markets; privatizing the banking system; facilitating information and transparency; developing clear rules of the game for savings, investments, and taxes; and making the legal system more efficient and credible.

Signs of Change

We began this chapter by noting that little is known about the relationship between capital markets, the environment, and the needs of future generations. Despite the need for a great deal more research, there are many signs that capital markets are beginning to play a much more positive role in altering the direction of investments and of resources toward sustainable development.

The growth of "green funds" shows that private investors are becoming more concerned. These funds, most active in the United States and the United Kingdom, account for only a tiny fraction of overall investment but have been growing steadily in size and appearing in other European countries, such as Norway and Germany. They received a considerable boost in Britain in 1989 when a survey by the financial group Eagle Star found that 72 percent of its unit trust holders rated the ecological stance of a fund manager as important.[14] They have also been helped by surveys showing that environmental mutual funds have regularly outperformed the Standard and Poor 500 index.[15]

The robust growth in emerging markets offers another sign of hope. David Gill, former director of the IFC capital markets department, has suggested that annual new investment in these markets could grow to $100 billion.[16]

The global integration of capital markets should improve the quality and quantity of information available everywhere, as well as corporate valuing practices.

Last but not least, there also remains the hope of a peace dividend, due not only to the end of the cold war but also to multilateral lending agencies' growing pressure on nations to invest more in human capital and sound resource management and far less on arms and armies.

This combination of behavioral and structural change suggests that capital markets may play a large role in redrawing the world economic and environmental map.

5   Trade and Sustainable Development

"If environmental progress is not to remain solely the property of affluent nations, developing nations must have their fair shot at progress. Free trade incorporating sound environmental principles enhances that prospect of advancement."

Jay D. Hair
President, National Wildlife Federation, United States

The ideal of "free trade" is today under attack from two different camps: those who would intervene for the sake of the environment, and those who are motivated to intervene on the basis of new theoretical concepts—reciprocal trade, negotiated trade, and managed trade.

We feel that it would be a great tragedy if the sustained efforts on the part of the General Agreement on Tariffs and Trade (GATT) suffered a reversal. It would be a tragedy laced with much irony. Traditionally, the industrial nations of North America and Europe have championed free trade, against the resistance of most developing nations and centrally planned economies. Today, it is the former that tend to question the benefits of liberalized trade, while developing nations and the newly emerged democracies of Eastern Europe see it as their main hope for economic development.

Unless nations trade, they cannot develop. Unless nations develop economically, they cannot protect their environments, clean up environmental damage, or make efficient use of resources. Of course, free trade—like democracy and other forms of free choice—has its dangers; freedom includes the freedom to make mistakes. But there appears to be no better system for allowing people and nations the maximum potential for making the right choices. Thus free trade has a role to play in progress toward sustainable development. This is a difficult truth for business because the great majority of the barriers to trade—tariff and other-wise—are the result of businesses lobbying governments. Given this, business must work with governments to bring down those barriers.

How, then, can free trade be made to support the internalizing of environmental costs, an integral part of the foundation of sustainable development? The best answer is that it cannot and should not. Inter-nalizing environmental costs and making polluters pay must remain the responsibility of individual governments. Trade cannot and should not be bent to serve any purpose other than allowing nations and companies to seek out and take advantage of comparative advantages.

It must be accepted that sovereign countries are free to choose the methods by which they assign values to their own environmental re-sources. International differences in the costs of conserving resources will be reflected automatically in each country's comparative advantage.

Goals such as environmental protection and sound resource manage-ment cannot be secured by unilateral trade measures. Furthermore, they are too important to be left to the volatile and ever-changing process of trade. The harmonization of environmental costs and values will be necessary only when the international environment is at stake. That can be achieved through the negotiation of international environmental agreements. Carefully drawn, these need not threaten trade or the trading system. If such treaties are not forthcoming, the pressure for tariffs will mount.

There are even situations in which more liberal trade would actually encourage the removal of market distortions that are harming the envi-ronment. For example, if industrial nations were more open to the import of agricultural produce from the developing world, Northern farmers would gradually cease to overuse chemicals, energy, and land to pro-duce crops such as sugar and rice, which tropical nations should be able to grow more efficiently.

Taking a long-term perspective, it follows then that economic growth, trade expansion, and environmental protection are goals that can only be reached in conjunction.

Trade Then and Now

The General Agreement on Tariffs and Trade, established in 1948, is a multilateral agreement with 103 member countries (as of October 1991).

Its primary goals are to encourage the liberalization of international trade and to foster economic and social growth. GATT expressly limits each contracting party's recourse to discretionary trade restrictions in order to create a more stable and predictable business climate for producers, investors, and traders, and hence to help guarantee the conditions for free and fair competition on world markets.

The GATT agreement operates through successive rounds of multilateral negotiations to reduce incrementally tariffs and the nontariff barriers to trade (such as import quotas). The Uruguay Round of negotiations, which dragged on well past its deadline at the end of 1990, aimed not only to liberalize world trade further but to extend GATT's reach to cover trade in services and matters related to intellectual property, as well as to sharpen its effectiveness in areas such as trade in agricultural goods.

Liberal trade is generally seen to have helped the industrial world to develop. "The productivity levels of the major industrial countries converged after the war mainly because of trade liberalization," notes a World Bank report. "This removal of major trade barriers facilitated the transfer of technology among industrial countries—mainly from the United States to Europe and Japan." The Bank adds that "lifting of the industrial countries' protectionist policies is critical to the developing countries' prospects for improving productivity levels."[1]

The volume of world trade grew by an average 3.5 percent a year during the 1980s, accelerating during the decade to reach a record $3.1 trillion in 1989, 7.5 percent above the level in 1988. One quarter of international trade is covered by bilateral agreements, another quarter is barter, a quarter is between subsidiaries of the same corporations (a figure large and rising due to the rapid increases in direct foreign investment), and a mere quarter is trade that might be termed multilateral.[2]

Three major developments are affecting the nature of trade in the 1990s: a rise in protectionism in the North accompanied by a decline in protectionism in the South; a sharp increase in direct foreign investments, which in a sense is trade by another means; and increasing conflicts, both ideological and concrete, between trade policies and environmental policies.

Between 1983 and 1991, the foreign direct investments of multinational corporations grew three times faster than world trade. Four fifths of the total was accounted for by the United States, Japan, and the European

Community.[3] To put this trend another way, while international trade volumes increased 5 percent a year between 1983 and 1988, global direct foreign investment grew by more than 20 percent annually in real terms.[4]

Trade and Environment

Environmental groups are becoming increasingly concerned about the effects of liberalized trade on the environment. Some of their concern is based on a rather simplistic view that economic growth necessarily hurts the environment, and trade must therefore be bad because it spurs economic growth. Others are worried that liberalized agricultural trade will encourage debt-strapped developing nations to clear forest and other natural areas to grow more export crops. Such groups cannot easily insert their concern into bilateral agreements, or barter agreements, or trade within one company; not surprisingly, then, they have tended to focus on GATT.

In some countries, environmental groups have allied themselves with others traditionally in favor of protectionism, such as farm organizations. Both want to limit trade in commodities, but for quite different reasons. This is an odd alliance between those interested saving the environment, who usually believe that poverty encourages environmental destruction, and groups interested in reducing the competitive advantage of developing countries, and thus end up keeping the poor impoverished.

The Organisation for Economic Co-operation and Development (OECD) has predicted that "with the development of the international dimensions in both trade and environmental issues, it appears that the potential for conflicts in the 1990s between trade and environmental objectives is on the rise."[5] To limit such conflicts, it is important to identify legitimate points of contact between trade and environment, and to distinguish environmental problems that extend beyond national boundaries. In practice, such a distinction is not always easy, but the concept gives valuable guidance on which policies are likely to be most effective. International environmental problems include the loss of biological diversity, the depletion of the ozone layer, and global warming. Solving these problems requires cooperation among sovereign states.

Restraints on international trade have figured in several recent environmental agreements. The Convention on the International Trade in

> *"In the context of world economic growth, sustainable development has a vital role to play in reminding businessmen that their business strategy should always incorporate environmental protection as a major principle."*
>
> Y.A.M. Tunku Naquiyuddin ibni Tuanku Ja'afar
> Chairman
> Antah Holdings Berhad

Endangered Species of Wild Flora and Fauna limits the species, and their parts, that may be traded internationally. The Montreal Protocol on Substances that Deplete the Ozone Layer requires parties to phase out trade in chlorofluorocarbons and other covered substances, and it threatens future trade restrictions on products containing or made using the controlled chemicals. And the Basel Convention on the Control of Transfrontier Movements of Hazardous Wastes and their Disposal limits trade directly by restricting the transboundary movements of certain materials under certain conditions. The International Tropical Timber Agreement is the first and so far only commodity trade agreement to contain environmental and conservation goals.

Such agreements can and should be made compatible with existing international trade rules. Their effectiveness should not depend on the threat of trade restrictions against countries that do not comply with international environment standards. That simply covers up the failure of cooperative negotiation. The international trading system is rule-based, not power-based, and it is important to business and governments that it stays that way.

The greatest danger to this system is not from international agreements but from the passage by individual nations of measures that affect trade for environmental purposes, either global, pending the negotiation of agreements, or, more probably, local. The U.S. government, for example, banned the import of tuna from Mexico, claiming that Mexican tuna fleets kill too many dolphins during their operations. The effect of this has been, the U.S. government maintains, to put pressure on Mexican fleets to kill fewer dolphins, but the Mexican government notes the effect is to reduce competition to U.S. tuna fleets. A GATT arbitration panel concluded in August 1991 that the U.S. action was contrary to the GATT agreement, as it represented an unreasonable and unjustifiable barrier to trade. The decision said in part that "a contracting party may not restrict imports of a product merely because it originates in a country with environmental policies different from its own."[6]

A similar case involves a Danish law requiring that beer and soft drinks be sold only in returnable bottles with a compulsory deposit. Brewers from other countries protested, and the case went before the European Court of Justice, based on the claim that the Danes were imposing disproportionate environmental protection. But the Court backed Denmark, saying that environmental considerations justified a minor constraint on trade so long as protection of domestic industries was not the goal of the legislation.[7]

Such issues can be dealt with in several obvious ways. One is for nations with similar environmental problems and at similar levels of development to work to harmonize voluntarily their approaches to environmental problems. Total harmonization is impossible. But if nations can agree on and all use, for example, techniques of catching tuna that catch few dolphins, then these issues need not muddle trade agreements. Another approach is to improve product labelling, and then rely more on consumer choice and less on legislation to keep out environmentally damaging trade goods.

Perhaps the most effective way forward is to improve the ability of GATT to minimize trade interference caused by environmental regulations. It is hardly surprising that GATT does not deal effectively with environment or sustainable development issues, as neither was an international concern when it was set up. Promoting more liberal trade should still be the primary objective of GATT. The organization can reduce the trade interference of national regulations by encouraging their harmonization. But GATT cannot be transformed into an agreement that puts more emphasis on environment than on trade.

Since 1965 GATT has had a Committee on Trade and Development concerned with the trade problems of the South. In 1972 it established a group on Environmental Measures and International Trade, following fears raised at the U.N. Conference on the Human Environment that environmental concerns would lead to trade barriers. As of late 1991, this latter group had not met.

Today the instruments available to GATT to deal with environmental issues are primarily its Article XX (which does not use the word environment) and its Agreement on Technical Barriers to Trade (which does use the word).

Article XX states that countries have the right to adopt trade measures "necessary to protect human, animal and plant life or health" and "relating to the conservation of exhaustible natural resources if such measures are made effective in conjunction with restrictions on domestic production or consumption." The measures taken must not result in "arbitrary or unjustifiable discrimination between countries where the same conditions prevail" and they must not represent "a disguised restriction on international trade."[8]

The Agreement on Technical Barriers to Trade provides a framework for dealing at the multilateral level with trade-related issues arising from technical regulations and standards. It encourages international harmonization of standards and recognizes explicitly that "no country should be prevented from taking measures necessary...for the protection of human, animal or plant life or health, or the environment...subject to the requirement that they are not applied in a manner which would constitute a means of arbitrary or unjustifiable discrimination between countries where the same conditions prevail as a disguised restriction on international trade."[9]

How should the GATT rules be changed so that the agreement can cope with interference between concerns for free trade and concerns for the environment? Before a precise answer can be given to this central question, several issues should be considered. First, Article XX should not be allowed to become a loophole to be used by protectionists. Breaching free trade principles should be seen as exceptional, and this exceptional character should also be maintained in the case of environmental danger. Second, negotiators should consider whether and how GATT's dispute settlement mechanism can be used to settle environmental conflicts arising through trade. This would help the enforcement of internationally negotiated environmental rules.

In addition, the following fundamental principles should be introduced into GATT law to address environmental issues:

• *Transparency*: Notification requirements need to be introduced so that all environmental regulations with potential trade impacts become internationally unambiguous. Today no international forum exists to which environmental regulations may be reported and where trade implications can be assessed. Such a forum needs to be created quickly, preferably under the auspices of GATT.

• *Legitimacy*: Environmental measures that restrict trade should be legitimate, and thus backed by strong scientific evidence. International bodies or panels of scientific experts should be established to test the legitimacy of such measures. But where environmental threats are particularly serious or irreversible, GATT should adopt the precautionary principle, erring on the side of prudence.

• *Proportionality*: Trade-restrictive measures should not go beyond what is absolutely necessary to produce a desired environmental result. Here, too, internationally accepted criteria and perhaps advisory panels could help.

• *Subsidiarity*: Every time an environmental goal can be achieved without a measure affecting trade, trade-related measures should be avoided.

GATT should respond quickly to the Group of Seven's call to define how trade measures can properly be used for environmental purposes. It should immediately establish a working party on introducing further rules aiming at harmonizing the requirements of free trade and those of environmental protection.

The organization should be consulted and should play an active role in the negotiation of future environmental agreements, such as the ongoing climate negotiations, that foresee the need to include trade-related provisions. Under the principle of subsidiarity, consideration should be given to whether trade measures are strictly essential for achieving a given environmental goal. If so, measures that distort trade and competition the least should be implemented. Including GATT in this way would not only ensure that trade-related issues are properly taken into account, it would also reduce GATT's direct involvement later in disputes caused by agreements that were not carefully prepared.

Trade and Development

It is clear that trade restrictions by all countries slow, and in some cases virtually abort, the economic development of developing countries. Increasing market access in both industrial and developing countries is a vital though by no means sufficient condition for development.

> *"Consumers in the highly industrialized countries are beginning to reject products derived from the destruction of the biodiversity of the tropical forests. And new markets are opening up for products produced respecting that biodiversity, such as wood produced from trees that are planted and hewn in a sustained development cycle."*
>
> Eliezer Batista da Silva
Chairman
Rio Doce International S.A.

It has been estimated that protectionism costs developing countries $100 billion in lost potential revenues for agricultural products, and a further $50 billion in lost potential textiles sales.[10] The proportion of industrial-country imports affected by nontariff barriers roughly doubled between 1966 and 1986, to more than 40 percent, according to somewhat controversial World Bank estimates. Yet this reflects the changing structure of exports from developing countries as well as new nontariff barriers.[11]

Also, developing nations, many of which have environmental advantages in their abilities to produce some crops, are competing against North American and European countries that heavily subsidize agriculture. In 1990, OECD countries spent $176 billion on farm subsidies.[12] Such spending—nearly three times their total official development assistance—not only keeps industrial-country produce artificially cheap, it creates large agricultural surpluses that have a distorting effect on world trade. They are often dumped on developing countries, whether these nations have food shortages or not, where they compete unfairly with local produce and dampen government enthusiasm for agricultural development.

Problems of market access are not confined to agriculture and textiles. The attempts of developing countries to diversify into higher value-added production are also hampered by lack of access to markets in industrial countries, thereby restricting growth and development. For example, many industrial countries have higher import duties on furniture than on raw timber. And their efforts to move away from primary exports toward more profitable items or those with faster growing world demand are impaired by the shortage of foreign exchange to buy inputs for the diversification work.

There has also been high protectionism among developing nations, especially those pursuing import substitution strategies and attempting to protect infant industries. While some of this protection may have been necessary, such strategies isolated the new industries in many countries from competition, and thus from innovation. Most developing countries are now moving away from import substitution toward export-led approaches, and are lowering artificially high currency values, as it becomes clear that building national competitive advantage is the only sure way of ensuring development.

Mexico got rid of many of its tariffs in 1986, a step credited with making Mexican industry more competitive. This new efficiency, and a weak peso, helped more than double total Mexican exports to the United States between 1985 and 1991. Mexico has also greatly increased its U.S. imports. According to U.S. professor of economics and international affairs Gene Grossman, "It's allowing each country to specialize, to take advantage of its strengths, and that has resulted in a more efficient use of resources, particularly in Mexico, because it's so much smaller."[13]

Of course, developing countries need more than free trade to thrive and to attract foreign investment. They need stable governments, sound macroeconomic policies, clear and open financial policies, vigorous education and training systems, and increased reliance on domestic savings to finance investment. (See chapter 10.)

Yet just as many nations are opening their borders to trade and investment, they are finding that Northern investors are putting money into other Northern countries. The U.N. Centre on Transnational Corporations described in 1991 a world of three major trading blocks, or "triads," in which private but global companies are shifting from an emphasis on exports from the home country to investing and manufacturing in other countries. It further found that the developing world received a quarter of these investments during the first half of the 1980s but only 18 percent during the second half. And about 75 percent of these funds went to just 10 countries: China, Hong Kong, Malaysia, Singapore, and Thailand in Asia; Argentina, Brazil, Colombia, and Mexico in Latin America; and Egypt. During the second half of the decade, the least developed countries received only 0.7 percent of the average annual investments going to the developing world.[14]

Trade and Sustainable Development

As clearly evident by now, a number of challenges must be faced in the transition to a world of freer trade and more sustainable paths of development.

Some industrial countries that once championed free trade—and whose leaders still promote it—are becoming more protectionist as they are becoming less competitive due to rising costs. Thus while demanding that developing nations restructure their economies to develop more efficient, more open markets, many industrial nations erect barriers against the participation of these restructured nations in the global marketplace.

Much is made of the relatively small amounts of development assistance that goes to developing countries. It is indeed unrealistic to expect present amounts and styles of aid to be much help in levelling the trade/ environment playing field among nations. Aid has stagnated over the 1980s and shows no real signs of increasing. And the worst forms of aid are little more than international market distortions. The problem of the dumping of agricultural surpluses has been mentioned. "Tied aid"— assistance given on the condition of purchases from the donor nation— often forces poor countries to buy materials of dubious value, or to pay more for useful goods than they would have in open competitive bidding. Many poor countries would be better off if industrial nations discontinued much aid and instead opened their markets. OECD official development assistance for 1990 totalled $62.6 billion, yet as noted earlier, developing countries lost at least $150 billion in revenues due to barriers to trade.[15]

Developing nations that rely largely on agricultural exports must take a hard look at their real costs. We have argued that free trade may encourage pricing that better reflects environmental costs in industrial nations accepting developing-country agricultural imports. But poor countries may be exporting a portion of their own "sustainability" if they are producing export crops by overusing soil, water, and forest systems and thus reducing future productivity. As Norwegian Nobel economics laureate Trygve Haavelmo and Stein Hansen note, "export is not an end in itself....Developing countries must realise they must stringently avoid exports that they cannot afford."[16]

The question that remains—and to which we have no convincing answer nor indeed know of one—is how can a developing nation charge a price for exports that reflects their environmental costs, and compete against other nations willing to absorb such costs for short-term profits. Obviously this will require international harmonization, but the nature of this process remains unclear. It may be helped by altering standardized national accounts (the basis of calculations of the gross national product) to include values of environmental goods and services, as discussed in chapter 2. This would not tell nations what to charge for exports, but it might help them identify those they could not afford over the long term.

But it is not the job of one nation to "punish" another for selling exports at less than their environmental costs. As noted in the beginning of this chapter, internalizing environmental costs is a domestic responsibility and also a goal to be moved toward internationally through methods other than trade. Otherwise, the result will be a chaos of unilateral nontariff barriers. Such obstacles, if imposed unilaterally in the North, allow Southern industries little flexibility of response, little in the way of a predictable or reliable regulation environment, and little in terms of gradual introduction. This would be precisely the sort of "environmental imperialism" feared by many developing nations. It would also destroy many of GATT's achievements of the past four decades.

Industrial nations are introducing increasing numbers of standards, regulations, and economic instruments to internalize environmental costs. If these are not coordinated through international negotiation, they will in many cases constitute nontariff barriers to trade. Some of this harmonization will take place in GATT; most of it will probably happen in other forums. Key issues in these negotiations will be how much time and what forms of help and motivation to give poorer countries to reach universal standards. And how can local, national, and regional differences be taken into account?

There is general agreement that developing countries should be allowed longer phase-in periods, according to their individual states of development. It would be neither fair nor economically sensible to force them to bear the full and immediate cost of an adjustment to stringent international environmental standards at a time when they face enormous development costs. In subscribing to restrictions they will adhere to later, in some cases much later, these countries should not be considered "free riders." And thus they should not be subject to the punitive trade measures included in some international agreements against free riders.

In December 1989, the OECD aid ministers noted: "The developed and other economically advanced countries cannot live in isolated enclaves of prosperity in a world where other countries face growing mass poverty, economic and financial instability and environmental degradation. Not only is this unacceptable on humanitarian grounds: the future well-being of developed countries is linked to economic progress, preservation of the environment and peace and stability in the developing world."[17]

The protection of the environment, trade expansion, economic growth, and development are complementary and interdependent parts of a strategy for sustainable development. Yet the danger of conflict between environmental regulation and further trade expansion is real and increasing. Such conflict needs to be minimized by awareness-building and by establishing clear rules of the game. Further progress toward harmonization of environmental regulations must be made.

6   Managing Corporate Change

"The world we have created today as a result of our thinking thus far has problems which cannot be solved by thinking the way we thought when we created them."

Albert Einstein

Sustainable development stands at the center of a global economic, technological, social, political, and cultural transformation that is redefining the boundaries of what is possible and what is desirable.

For business, this means profound change: change in the goals and assumptions that drive corporate activities, and change in daily practices and tools. Continued economic development now depends on radical improvement in the interactions between business and the environment. This can only be achieved by a break with "business as usual" mentalities and conventional wisdom, which sideline environmental and human concerns.

The challenge of managing corporate change to move forward simultaneously in both economic development and environmental protection is immense, but we are not starting from zero. Business has proved its ability to manage fundamental changes in planning and action in the "quality revolution," which has influenced virtually every business in every industry throughout the world. This recent revolution has also demonstrated business's ability to move in tandem toward objectives that seemed to be opposed—in this case, increasing quality while lowering costs.

Many of the facts, data, and examples in this chapter were provided by BCSD members and their colleagues, and thus are not referenced.

The tools and processes used in this revolution, which are still being developed, along with the experience gained and the results produced, provide a foundation on which business leaders may build toward a sustainable future. In fact, many firms that have embraced total quality management as a comprehensive management philosophy see environmental excellence as a natural extension of it.

The Emerging Context

The old and still prevailing view of the links between business and the environment is that environmental protection and profitability are natural opposites. Improving the environment is thought to mean reduced profitability for business and increased costs for consumers, while profitability is thought to require environmental consumption and degradation.

In other words, it is often assumed that we can have either a healthy environment or a healthy business sector, and that wise governance requires making suitable trade-offs so that these polar opposites are kept in proper balance. It comes as no surprise, then, that responses to pressures about environmental quality have often been reactive and involuntary, largely defined by others through laws, regulations, and consumer pressure. In general, during the past 20 years business has tended to be overcautious and conservative in its approach to these challenges, underestimating the possibilities for positive change.

Society can no longer afford this. It is time for businesses to take the lead, because the control of change by business is less painful, more efficient, and cheaper for consumers, for governments, and for businesses themselves. By living up to its capabilities to the full, business will be able to shape a reasonable and appropriate path toward sustainable development.

Thus sustainable development means moving business decisions toward both a healthy environment and a healthy economy at the same time. This suggests that environment and economy are intertwined, not as enemies but rather as partners, in the global pursuit of an enhanced quality of life.

This chapter offers a model of corporate commitment and change management. It is embedded in the context of the new assumptions that capture more adequately society's emerging needs. The role of leadership is to focus that context through the articulation of a corporate vision. Leaders know that for a new context to emerge, deeply ingrained assumptions must be brought forward, tested, and modified.

Any vision of sustainable development is dynamic, and must be open to revision as it is realized. Clear and committed vision leads to the development of strategies and actions that change corporate processes and systems in ways that align them with the new vision. The proof that the vision is becoming reality lies in the measurement and reporting of its results or outcomes. These results are part of a feedback loop, stimulating continuous improvement between corporate process and corporate outcomes. In this way, the corporation moves to higher planes of competitiveness and efficiency.

The Vision of Sustainable Development

An increasing number of corporate leaders are convinced that it makes good business sense to secure the future of their corporations by integrating the principles of sustainable development into all their operations in order to:

• recognize that there can be no long-term economic growth unless it is environmentally sustainable;

• confirm that products, services, and processes must all contribute to a sustainable world;

• maintain credibility with society, which is necessary to sustain business operations;

• create open dialogue with stakeholders, thereby identifying problems and opportunities as well as building credibility through their responses;

• provide meaning for employees beyond salaries, which results in the development of capabilities and growth in productivity; and

• maintain entrepreneurial freedom through voluntary initiatives rather than regulatory coercion.

A clear, committed corporate vision transforms the everyday context within which employees see and do their jobs. According to Jérôme Monod, chairman and chief executive officer (CEO) of Lyonnaise des Eaux-Dumez, "as with all industrial revolutions, that of sustainable development must be accompanied by a cultural revolution within the company." In other words, a vision unfolds in the context of corporate culture.

The issue is not whether the company vision looks good on paper, but whether behavior and outputs change. When vision is understood to be a common context for action, as opposed to just a "big goal" or an idealized picture of the future, it provides the framework and guidelines for stimulating action. Rather than a project to be completed so we can move on to something else, continuously revising and focussing the vision becomes a major role of top management.

Senior management has the authority and perspective to gather the input from stakeholders and monitor results, and to use these as foundations on which to base a company's strategic direction in order to promote a sustainable development consciousness throughout the organization. Only firm leadership from top management can reconcile the goals of long-term sustainability and short-term profitability.

In January 1990, for example, Monsanto CEO Richard Mahoney set a goal of cutting air emissions of certain hazardous chemicals 90 percent by 1992. Most of the company's technical people said it could not be done; yet a year later Monsanto was well on the way to meeting its goal. New knowledge and new technologies have been developed, and staff are enthusiastically working to make further breakthroughs.

Organizational commitment is needed to integrate environmental aspects into all activities—from research and development to production and distribution. Strategies and action plans are required to change the corporation's processes and systems to align them with a corporate vision incorporating the principles of sustainable development. Companies will need to review existing product portfolios and production systems and choose which to keep, which to update, and which to scrap. Responsibilities at the board of director, management, and operating levels must be clearly understood and reinforced by appropriate incentives.

Although individual leaders can make a difference in changing the framework of goals within which a company operates, they cannot transform this into a living reality without a critical mass of other committed individuals. When people share a vision and commit themselves to act on it, breakthroughs are not only possible, they are in time inevitable.

When viewed within the context of sustainable development, environmental concerns become not just a cost of doing business, but a potent source of competitive advantage. Enterprises that embrace the concept can effectively realize the advantages: more efficient processes, improvements in productivity, lower costs of compliance, and new strategic market opportunities. Such businesses may expect to reap advantages over their competitors who lack vision. Companies that fail to change can expect to become obsolete.

In sum, when a new context is created, everything changes. New questions can be asked, new choices appear, new processes are designed, and prior constraints can become the stepping stones for building a sustainable future.

Mobilizing the Vision Through Stakeholder Partnerships

Business is being challenged by a much broader and more diverse group of people who have a stake in corporate actions than ever before. Broadly, these "stakeholders" include not only customers, employees, and shareholders, but also suppliers, government, neighbors, citizens' groups, and the public. Involving these people, with all their differing views and concerns, usually leads to better decisions and more universal support for their implementation.

Considering stakeholder involvement to be legitimate and strategically important requires more effort than traditional public relations or information-sharing responses. New forms of collaboration are needed, including focus groups, advisory panels, forums for dialogue, and joint ventures. Building stakeholder involvement in the context of sustainable development extends the idea of corporate responsibility in time and space. Companies now have to consider the effects of their actions on future generations and on people in other parts of the world. Prosperous companies in a sustainable world will be those that are better than their competitors at "adding value" for all their stakeholders, not just for customers and investors.

Successful stakeholder involvement on sustainable development issues has been shown to have many benefits:

• *Employee support and personal responsibility:*   Employee involvement is a key aspect of Dow Chemical's Waste Reduction Always Pays (WRAP)

> *"If we view sustainable development as an opportunity for growth and not as prohibitive, industry can shape a new social and ethical framework for assessing our relationship with our environment and each other. Seeking cooperative dialogues and partnerships among and between industry, government, special interest groups, and the public will move the process of sustainable development much further, much faster."*
>
> Frank Popoff
> President &
> Chief Executive Officer
> The Dow Chemical Company

program. Staff are rewarded for their work on waste reduction projects each year. Not only is WRAP designed to reduce waste, it is also intended to broaden employees' thinking and instill a sense of responsibility, so that they never assume that someone else in the organization is looking after their wastes for them.[1]

• *Improved morale:*   Following a spill of acid wastes at Rowe Manufacturing in the United States, the United Auto Workers union launched a cleanup campaign, cutting in half the amount of chemicals used and substituting soap and water for many toxic solvents. Rowe has since formed a joint health and safety committee with representatives from both the union and management. Rowe vice-president John Nigro has welcomed the active cooperation of the unions, saying "there's been a significant improvement in the morale of workers."[2]

• *Better policy advice:*   Canada has perhaps the longest history of building a national coalition to support sustainable development. Responding to the World Commission on Environment and Development, the government established a National Task Force on Environment and Economy, with seven Ministers, six CEOs from some of Canada's largest corporations, and representatives of the environmental and academic communities. Many of the CEOs became enthusiastic advocates of sustainable development. Their excitement encouraged national business associations such as the Business Council on National Issues, the Conference Board of Canada, and the Chamber of Commerce to organize programs on sustainable development.

Moreover, CEOs of both large and small companies play active roles in the round tables on environment and economy that were subsequently established at the national and provincial levels. These round tables have become strong examples of multiple stakeholder cooperation and commitment to sustainable development.

Several Canadian firms are using advisory panels to facilitate dialogue with various public interests. For example, in 1989 TransAlta Utilities established a panel consisting of 13 members representing a cross-section of interests, including education, health care, environment, industry, farming, fish and game, law, and consumers. The group helped TransAlta develop its environmental mission and policy statements.

• *Public acceptance of corporate activity:* In the United States, McDonalds has formed a joint venture with the Environmental Defense Fund to develop a more efficient and more publicly acceptable waste management strategy. The U.K.-based International Institute for Environment and Development advises Shell on its forestry plantation work in Southeast Asia, particularly on ways to improve the livelihood of people living in and around the plantations.[3]

• *Reduced risk and liability:* Suppliers are increasingly being required to guarantee the environmental quality of their components and processes, as companies cannot afford to do business with a supplier who presents large-scale environmental risks. For example, at the U.S. products firm S.C. Johnson, an existing supplier relationship program was broadened to include environmental issues.[4]

• *Self-regulation rather than legislation:* Some governments are using voluntary agreements or covenants to reduce industry's environmental impacts. Such arrangements are mutually beneficial: for government, a covenant can be introduced far faster than legislation; for business, covenants allow far greater control over the implementation of agreed environmental targets. The Netherlands National Environmental Policy Plan led to two important covenants—one to cut pollution from the electricity generation sector, and the other to reduce packaging waste and encourage recycling.[5]

The U.S. Environmental Protection Agency (EPA) has launched an initiative in favor of self-regulation within U.S. business. Called the 33/50 Program, it aims to win voluntary commitments from 600 companies to reduce their 1988 emission levels of 17 high-priority toxic pollutants by 33 percent by 1992, and by 50 percent by 1995. By mid-1991, EPA had received pledges from more than 300 companies promising to eliminate more than 90 million kilograms of the targeted pollutants.[6]

Japan's Ministry for International Trade and Industry launched a "New Earth 21" program to tackle the threat of global warming by designing and producing new technologies to lower dependence on fossil fuels. It has won wide support from the Japanese industrial sector.[7]

• *More certainty from regulators:* In May 1991 the U.K. government established an Advisory Committee on Business and the Environment (ACBE), headed by John Collins, chairman of Shell UK. The ACBE provides the government with strategic guidance and workable proposals on how business can contribute to sustainable development. The fact that ACBE has been set up jointly by both the environment and the trade and industry departments shows the increasing integration of environmental considerations in government decision making, which should result in clearer and more effective signals for the business community.

In sum, the inclusion of stakeholders is a complex and time-consuming process with important benefits. Especially important is the recognition of government as a stakeholder and partner. The breakthrough required is inventing new ways to involve stakeholders so that decisions can be timely and enduring.

New Markets, New Management Strategies

Stakeholder involvement is only one aspect of effective corporate leadership in sustainable development. Commitment to a vision must translate into strategies and action plans. This often involves reorganizing, restructuring, and redesigning many processes and detailed systems within a corporation.

Strategies to reposition a firm for sustainable development could involve some painful choices for companies in "sunset" sectors. The producers of lead for gasoline, phosphates for detergents, and chlorofluorocarbons, for example, have all seen their businesses shrink because of environmental imperatives. But such companies need not face the same fate as the materials they provided if they can develop substitute materials that meet consumers' needs at reduced environmental costs.

Consider the asbestos industry. Asbestos cement was once seen as the ideal construction material—cheap, versatile, resistant, and usable in all types of buildings. During the 1970s, however, evidence was growing that asbestos fibers, when inhaled, present a serious health risk. As scientists were unable to indicate a safe threshold of exposure to asbestos dust, in 1980 the Swiss Eternit Group decided to replace asbestos cement by an as-yet-undeveloped material called fiber cement by 1990 at the latest, while phasing asbestos out altogether. In a far-reaching technology cooperation program between the Swiss parent and its affiliates, it turned the problem into an advantage through innovation. In some countries, fiber cement could be produced with locally available renewable materials at lower cost and was able to gain considerable market shares, since the new products were better adapted to local needs.[8]

Sometimes strategies for sustainable development involve turning old constraints into new opportunities. For example, at the U.S. utility New England Electric, senior management knew the appeal of conservation to many of the stakeholders. Yet conservation by its very nature erodes electricity sales, the core of New England Electric's business. By working with stakeholders to design an incentive structure for conservation, management was able to turn conservation into a profitable business opportunity.[9]

In some cases, companies have found new business opportunities by applying corporate assets to a new service, such as environmental protection. The environment industry is expected to become one of the fastest growing economic sectors. The international market is already valued at $280 billion annually, with potential to double by the end of this decade.[10] In 1990, the French water company Lyonnaise des Eaux and the construction group Dumez merged to form the world's largest company devoted to environmental management. Berzelius Umwelt Service became Germany's first publicly quoted company exclusively devoted to environmental protection, aiming to benefit from stricter waste disposal regulations and a decline in landfill capacity. And the Japanese government sees environmental technology becoming a boom industry of the future, with Japanese business set to reap the benefits of being ready to take advantage of this development.

Several other trends in our contemporary management culture will continue to evolve in the context of sustainable development. They concern the roles and responsibilities of the boards of directors and top management, the nature of organizational structures, and the training needs of middle management.

First, the traditional roles and responsibilities of boards of directors and top management are evolving to integrate the internal and external dimensions of a business and to provide new vehicles for stakeholder participation. To accomplish this, many companies have sought to establish the visible commitment of the CEO, the development and dissemination of corporatewide environmental policies, and the active involvement of their boards.

Some European firms, for example, have introduced a director with board-level responsibility for environmental matters; others have established a high-level position covering all environmental activities in the firm, such as the vice-president for environment at Norsk Hydro and the member of the management committee at Volkswagen. They can act as environmental champions in corporate decision making, and are usually supported by an environment department, providing policy advice and monitoring performance.

North American firms have tended to establish environmental subcommittees chaired by senior vice-presidents, or broadened the responsibilities of an existing subcommittee to incorporate environment concerns. And a few U.S. corporations have brought individuals with environmental expertise onto their boards.

Second, organizational structures are evolving toward broad network designs that are being stimulated by advancing communications and computer technology. Traditional linear and compartmentalized approaches are insufficient. A growing number of companies are drawing up policy statements as part of their commitment to environmental self-regulation. These have proved particularly important in large, decentralized multinational organizations as part of the creation of a corporate culture of sustainable development.

The task of integrating environmental policy into various departments is frequently facilitated through high-level coordinating committees. At Sony in Japan, for instance, a Global Environmental Council was formed in late 1990. It has four subcommittees that work on technology, products, resource and energy conservation, and postconsumer waste. In 1990, Mitsubishi Corporation established an environmental affairs department dedicated to promoting environmental awareness within the

> *"It has been argued that one cannot serve both the needs of industry and of the environment. I believe that this is not an impossible task. Industry can no longer afford to ignore environmental needs. Profit becomes pointless without quality of life. Financial accounts tell many stories but not all, and measuring performance by profit alone will not suffice. However, a greener future will remain an idealistic dream unless industrialists and environmentalists meet to transform it into a reality by talking and sharing problems."*
>
> J.M.K. Martin Laing
> Chairman
> John Laing plc

group worldwide and contributing to environmental protection projects. One of the department's first tasks was to support a tropical reforestation project in Malaysia.[11]

The corporate strategic program of Ciba-Geigy, Vision 2000, illustrates the integration of a vision of sustainable development into top-level efforts to direct and restructure a large corporation. In 1989, senior managers began a series of exploratory workshops to reposition Ciba-Geigy for long-term sustainable growth. The result was a deceptively simple vision statement: "By striking a balance between our economic, social, and environmental responsibilities we want to ensure the prosperity of our enterprise beyond the year 2000." The integration of the corporation's threefold responsibilities is essential.

Vision 2000 intends to develop a new relationship between "enlightened and determined leadership" at corporate headquarters and increasingly decentralized business units that have authority and responsibility. The number of management layers has been reduced, modern forms of teamwork have been introduced, and efforts have been made to encourage individual initiative and risk taking. Accountability for meeting economic, environmental, and social performance goals will more and more often belong with the heads of business units, who are setting their own targets. For example, the agricultural products division has set a goal of cutting production wastes by 50 percent, with corporate environmental staff providing technological support.[12]

The third evolving trend in contemporary management culture involves middle management. Roles are continuing to shift toward coaching individuals in their decision making so that real breakthroughs in business process and results are achieved. This requires new ways of thinking and the development of an organizational learning culture. Such a culture is based on an appreciation of the need to constantly rethink and be open to relearning the fundamentals of every aspect of business. The ability to tolerate uncertainty, design new strategies, coach, and use statistical tools for managing processes are other skills required to manage change for sustainable development.

Central to the concept of a learning organization is the view that the future, though it cannot be predicted, can be shaped. For almost two decades, the Shell Group planners in London have experimented with various methods, including scenario planning, to foster strategic thinking. Over time they have learned that what is important is not whether any scenarios come to pass (although the scenarios presented are plausible futures), but how the exercises help managers improve their mental models of reality. Managers are thus better prepared to cope with unexpected change, since the exercise of visiting plausible futures promotes flexible organizational responses.

Corporate Outcomes and Reporting

A vision and implementation strategy for sustainable development is most useful if the resultant actions can be reported, thereby offering chances for feedback and improvement. The basis for reporting is goals. Operational policies need to lay down corporate objectives and standards in ways that can be measured, for as the American business adage has it, "what gets measured gets done."

Some firms set down only general principles; others specify procedures such as internal environmental audits or performance reviews. What distinguishes policies is the degree to which they go beyond compliance with all government regulations, commit the company to minimal or zero pollution, apply the same principles or standards globally, engage in partnerships with governments and citizens, and adopt a life-cycle approach to product development. A clear corporate policy can provide a useful benchmark against which a business can assess its environmental performance.

A systematic, documented, periodic, and objective evaluation is needed of how well the organization is performing in the area of sustainable development, not only to facilitate management control practices but also to assess compliance with company policies, including meeting regulatory requirements. Sustainable development reporting is a demanding concept and goes beyond environmental reporting. It requires that companies assess their performance in both the environment and the economy in terms of quality of life today and for future generations. This goal has yet to be achieved.

Ideally, companies need to issue regular reports to their stakeholders on their progress toward sustainable development, for reaction and feedback. By receiving information that is standardized in content and form, readers would be able to evaluate whether a business is operating in a sustainable manner, its degree of improvement over time, and how its performance compares with others. The optimal report would thus be relevant to the needs and expectations of stakeholders, understandable without being vague, representative of business performance, and comparable over time and between companies.

In practice, most companies do not currently provide public information on environmental or sustainable development policies or activities, according to a survey of businesses in more than 30 countries commissioned by the BCSD and the International Institute for Sustainable Development, and conducted by Deloitte & Touche. Some businesses are experimenting with special types of reports designed for particular stakeholder groups, such as employees. Others are providing general purpose accounts of environmental activities that are distributed to all stakeholder groups. Still others are including in their annual reports separate sections on environmental issues and sustainable development.[13]

Many companies are developing internal reporting systems. Some ask line managers to report regularly whether they have achieved environmental and sustainable development targets. In some cases, senior management provides the board of directors with periodic statements combining financial, technical, and qualitative information.

Although agreed-upon measurement tools and widespread professional expertise do not yet exist, business is being pressured to take the lead in setting standards and providing supplemental reports for their stakeholders. Government regulations and the introduction of economic instruments that require companies to gather additional information anyway will provide baseline data that can be drawn on.

As a result of an initiative by the BCSD, the International Organization for Standardization has set up a Strategic Advisory Group on the Environment (SAGE) that is studying how to extend international standards beyond product quality, to measure, in other words, eco-efficiency. It is expected that international standards will be prepared in several fields, including methods to assess environmental performance, life-cycle analyses, environmental auditing, and eco-labelling. SAGE has found wide support within the business community.

Business has many good reasons to become proactive in its attitude toward reporting: customer pressure, community pressure (including that exerted in or on behalf of developing countries), pressure from shareholders and institutional investors, pressure from leading corporations, concerns about legal liability of outside directors (though liability concerns can inhibit the desire to report), emerging professional standards, and enforcement of environmentally sound behavior by international aid agencies.

The BCSD accepts as a baseline the International Chamber of Commerce principle of "compliance and reporting" that is to encourage business "to measure environmental performance; to conduct regular environmental audits and assessments of compliance with company requirements, legal requirements and these principles; and periodically to provide appropriate information to the Board of Directors, shareholders, employees, the authorities and the public." The term environmental audit describes "a management tool comprising a systematic, documented, periodic and objective evaluation of how well environmental organization, management and equipment are performing with the aim of helping to safeguard the environment by: facilitating management control of environmental practices; assessing compliance with company policies, which would include meeting regulatory requirements."[14]

In sum, the state of the art is evolving and must continue to evolve, for a decision to act without commitment to measurement is hollow. The specific outcome is less important than the rigor with which a company pledges itself to the validity of corporate reporting as an accurate reflection of its commitment to sustainable development.

The Leadership Agenda

Many CEOs have already taken a stand for sustainable development; they sense the urgency and are willing to take a strong leadership role.

The men and women at the forefront of sustainable development consider that their primary role is to ensure that the future for all stakeholders includes both a strong economic foundation and a healthy natural environment. Building a sustainable future depends on our absolute commitment to both. Many of the waste reduction and environmentally positive programs in business are economically viable and are providing positive rates of return in relatively short time periods. Yet there is much left to do.

We have at least the initial competencies, mechanisms, and resources available. What is needed is to focus and apply them in the context of a sustainable future. Corporate experiences in total quality management are directly applicable to the kinds of changes required to attain sustainable development. Our ability to forge effective relationships with all stakeholders, including governments, will allow the market to be engaged and our creative talents to be used in husbanding environmental resources.

Ultimately, sustainable development will result from the billions of choices individuals make every day. It will happen when people are committed to making it happen. And the time to make it happen has arrived.

7   The Innovation Process

"Many essential human needs can be met only through goods and services provided by industry....Industry has the power to enhance or degrade the environment; it invariably does both."

World Commission on Environment
and Development, 1987

During the past 20 years, a new global awareness has emerged that provides the basis for the shift to sustainable production and consumption patterns.[1] In place of the postwar era's stress on sheer quantity—"mass production" and "mass consumption"—greater emphasis is now being placed on quality; value-added is increasingly based more on knowledge than on resources or labor. Products and services are often customized to meet the desires of smaller groups of consumers.

In industrial countries, this period has been marked by the relative "dematerialization" of economic activity, seen most clearly in the decoupling of energy consumption from production growth following the two oil price shocks. Higher energy prices, combined with a drive for efficiency improvements, have meant that while the output of the chemicals industry has more than doubled since 1970, for example, its energy consumption per unit of production has fallen by 57 percent.[2]

Furthermore, the combination of ever more efficient resource use and tightening environmental regulation has significantly reduced certain types of pollution. In West Germany, the chemical industry managed to cut emissions of heavy metals by 60–90 percent between 1970 and 1987 while boosting output by 50 percent.[3]

Many of the facts, data, and examples in this chapter were provided by BCSD members and their colleagues, and thus are not referenced.

These improvements have been matched on an individual company level, as growing numbers of companies are regularly raising their "environmental efficiency"—the ratio of resource inputs and waste outputs to final product. At Nippon Steel Corporation, producing a metric ton of steel in 1987 emitted 75 percent less sulfur oxides and 90 percent less dust than in 1970. Since 1960, Dow Chemical has cut the production of hazardous wastes from 1 kilogram (kg) per kg of salable product to 1 kg per 1,000 kgs.[4]

At Ciba-Geigy, finished products represented only 30 percent of all outputs in 1979, the rest being waste. By 1988, the company's efficiency had increased to 62 percent; a goal of 75-percent efficiency has been set for the end of this decade.[5] In one particular process at Ciba-Geigy, producing 1 metric ton of a chemical called amide traditionally required 3 tons of highly corrosive phosphorous trichloride and 12 tons of water; 14 tons of resulting effluent had to be treated. This has now been replaced by a system using only 1.9 tons of raw materials and no water; the by-products are 0.6 tons of pure acetic acid, which can be recycled in other processes, and 0.3 tons of solid organic waste, which is incinerated.[6]

Nevertheless, overall waste and pollution emissions from industry in Northern nations continue to increase, outpacing economic growth. In France, 1 percent of economic growth currently generates 2 percent extra waste.[7] The U.S. Environment Protection Agency estimates that in the United States the generation of hazardous wastes is growing at an annual rate of 7.5 percent.[8]

New issues are emerging as the global economy enters the so-called postindustrial era. In the industrial world, the relative success in reducing pollution from factories is turning business's attention toward improving the environmental performance of products. The shift from mass production to "mass customization" has greatly increased the number and variety of products, posing new challenges: "if present trends continue, 50 percent of the products that will be used in 15 years' time do not yet exist," according to an Organisation for Economic Co-operation and Development (OECD) report.[9]

Under the pressure of tightening regulations, increasingly "green" consumer expectations, and new management attitudes toward extended corporate responsibility, companies are recognizing that environmental management now requires the minimization of risks and impacts throughout a product's life cycle, from "cradle to grave." This is in turn leading to the industrial ideal of an economic system based on "reconsumption"—that is, the ability to use and reuse goods in whole or in part over several generations.[10]

Companies now have to work with governments to spread environmentally efficient production processes throughout the global business community, paying particular attention to the needs of small and medium-sized enterprises and developing countries. This will require significant technological, managerial, and organizational changes, new investments, and new product lines. But as public policy continues to stimulate innovation toward industrial sustainability, it will be increasingly in a company's own interests to develop cleaner products and processes.

Cleaner Processes Through Pollution Prevention

All natural and industrial processes produce waste. Waste becomes pollution when it exceeds the carrying capacity of the environment, something that varies enormously over the vast range of materials and processes used and the differing ecosystems affected.

The commonsense, precautionary response to burgeoning pollution problems is to seek to prevent pollution before it happens. Where it is already occurring, the aim should be to eliminate the source of the problem rather than attack symptoms through often expensive "end of pipe" methods such as filters, scrubbers, treatment plants, and incineration.

Since the 1970s, this approach has been gaining ground in policy and corporate circles as the most cost-effective way of achieving environmental and economic efficiency. In the United States, a stated purpose of the National Environmental Policy Act of 1969 was "to promote efforts which will prevent or eliminate damage to the environment and biosphere."[11] More and more companies are realizing that the pollution they produce is a sign of inefficiency, and that waste reflects raw materials not sold in final products.

A combination of increasing regulatory pressures, mounting public expectations, and tightening competitive conditions is now driving companies everywhere to adopt the logic of pollution prevention. One powerful stimulant in the OECD region has been the growing costs of waste disposal: waste processing can now cost companies an average of

$380 per ton, rising to $3,000–10,000 per ton for toxic and hazardous wastes.[12] In addition, governments are starting to consider comprehensive programs to make the polluter pay through the sorts of liability and pricing policies discussed in chapter 2.

The 3M Company in the United States has pioneered corporate pollution prevention since 1975 with its Pollution Prevention Pays (3P) program. A multinational with total sales of over $13 billion in 1990, 3M has run more than 3,000 3P projects in the past 15 years, cutting air pollutants by 120,000 tons, wastewater by 1 billion gallons, and solid waste by 410,000 tons. In the process, the company managed to save $537 million.[13]

Experience has shown that the main barriers to pollution prevention are a lack of information, desire, and appropriate incentives. Both large and small companies can prevent waste and pollution, and thus save valuable raw materials, cut waste management costs, reduce liability, improve productivity, and thereby promote a more efficient allocation of corporate resources. For example, the Chromolux electroplating company in the Netherlands, which employs only 30 workers, found through a collaborative effort with local researchers that it could switch to a cyanide-free system, cutting waste by 60 percent. Chromolux made a total annual saving of Fl164,000 ($92,600) through raw material savings and reduced waste treatment costs.[14]

Many national and sectoral studies have found that industry has great potential to improve its environmental and economic efficiency even further through prevention. In 1988, the Dutch Technology Assessment Organisation studied the possibilities for pollution prevention in 10 typical companies, ranging from small firms to subsidiaries of multinationals. Known as PRISMA, the study concluded that industry can reduce some 30–60 percent of its pollution by preventing waste and emissions, while remaining competitive, by using existing management techniques and current technology.[15] In the United States, the Office of Technology Assessment estimated in 1986 that half of all industrial wastes could be prevented using existing technologies.[16]

The potential for energy, water, and materials conservation is broad: in the textiles industry, water conservation practices can often reduce water use by a quarter, according to a U.N. report.[17]

By adopting a pollution prevention approach, companies can start to take control of the process of environmental change in ways that make economic and operating sense, rather than seeing their own processes controlled by tightening regulations and expectations. In the United States, Monsanto pledged to cut air emissions of certain hazardous chemicals by 90 percent by 1992, en route to zero emissions, while 3M set itself the target of a 70-percent reduction of air emissions by the following year.[18]

Thus environmental considerations must be fully integrated into the heart of the production process, affecting the choice of raw materials, operating procedures, technology, and human resources. Pollution prevention means that environmental efficiency becomes, like profitability, a cross-functional issue that everyone is involved in promoting.

Pollution Prevention in Practice

The myriad of pollution prevention possibilities can be divided into four main categories: good housekeeping, materials substitution, manufacturing modifications, and resource recovery. Often companies employ a number of these approaches simultaneously to resolve a particular problem.

Good Housekeeping

The aim of good housekeeping is to operate machinery and production systems in the most efficient manner. As such, it is a basic task of management. For example, the proper operation and regular maintenance of equipment can often substantially reduce leakage and overuse of materials. Improvements in housekeeping practices, which can often reduce pollution by between a quarter and a third, usually do not require large capital expenditures.

Christopher Hampson, environmental director at ICI, admits that a quarter of that company's environmental costs come from "losses in containment and less than optimum operation of plant."[19] In the Hong Kong textiles industry, environmental improvements have been achieved through relatively simple good housekeeping practices. The volume of wastewater can be substantially reduced by such obvious practices as shutting off of water supply to equipment not in use, installing automatic shut-off valves in hoses, and supplying only the optimum amount of water to the machine.

Good housekeeping requires attention to detail and constant monitoring of raw material flows and impacts. Many companies still have no idea how much or what type of wastes and pollution they produce. Waste minimization and pollution prevention start from the basis of accurate measurement, identifying and then separating wastes. Improvements in information technology have also made environmental monitoring more affordable. Some companies have introduced sophisticated waste measuring and tracking systems. Du Pont, for example, recently installed an Environmental Data Management System at each of its facilities.

Waste can also be prevented through more efficient inventory control. Surplus raw materials that are no longer needed can be sold to third parties. Many Japanese companies have strong investment and resource recovery units that contribute to good housekeeping and pollution prevention.

Materials Substitution

Identifying and eliminating sources of pollution often implies restructuring for both producers and consumers. Full or partial phaseouts of lead, mercury, DDT, and chlorofluorocarbons (CFCs) have been implemented in various parts of the world as the only effective ways of solving the problems they cause.

Substituting one material for another offers the prospect of completely eliminating a given pollution problem. One of today's biggest issues for companies is how to deal with emissions of volatile organic compounds (VOCs), which are linked to two of the industrial world's major air pollution problems—photochemical smog and global warming. A major industrial source of VOCs is the use of solvents, particularly in the paint and coatings industry.

Car manufacturers are currently in the throes of switching from solvent-based to water-based paints. The German automaker Volkswagen has applied its "three Vs" environmental policy to the issue (Vermeiden, Verringern, Verwerten, or Prevention, Reduction, Recycling). Faced with Germany's new air pollution regulations, the company realized it could either apply end-of-pipe filters to existing technology or move to a new paint system, and thus reduce solvent emissions at the source. The company also wanted to improve the working environment, while increasing flexibility and paint quality.

Volkswagen decided to invest DM1.7 billion ($1.1 billion) in new paint shops at its Wolfsburg, Hannover, and Emden plants. As Rudolf Stobbe, environmental manager for the Wolfsburg plant, said, "We had to look to the future, and make a big leap, as you can't build a paint shop in steps." When it is completed in 1993, the new paint shop will use water-based paints for the base coat, cutting the solvent content of paint from 80 percent to about 10 percent. Volkswagen hopes to be able to move to water-based top-coat paints by 1996, thereby further reducing solvent emissions.

At the Swedish automaker Volvo, Sigvard Hoggren, vice-president for environmental affairs, explains his company's belief in a preventive solution: "In the long run, we must use materials in our processes that do not give rise to hazardous emissions at all."[20] By 1993, Volvo's main Torslanda plant aims to cut 1987 levels of solvent emissions by 75 percent by building a new paint shop that can use water-based paints.

Thousands of smaller firms also use processes that lead to VOC emissions. Often they need outside assistance to speed the transfer to cleaner processes. In Sweden, researchers from the University of Lund helped seven local firms to reduce pollution at the source. One, a 350-person electrical light fixture manufacturer, produced about 200 tons of VOCs per year in a metal coating process. Under government pressure to reduce these emissions, the company bought an end-of-pipe facility to incinerate the solvents. The system has never been used, however, because the Lund researchers offered an alternative that eliminated the VOC problem—a switch to powder paints. The investment in new paint equipment was paid back within the first year, and large savings can be expected in the future as the company avoids the operating costs of a combustion system. Energy efficiency and product quality have also been improved.[21]

Business is also seeking to phase out ozone-depleting substances such as CFCs. Many companies are finding that substituting other materials can be relatively straightforward and profitable in some processes. In the electronics industry, many companies have started using simple water-based solutions to clean printed circuit boards. IBM has pledged a worldwide CFC phaseout in electronics production by 1993.[22]

Manufacturing Modifications

Often companies can considerably reduce emissions by simplifying production technology through lowering the number of process stages.

Switching to closed-loop processing can also conserve resources and cut noxious emissions. This was the case at a film developing unit at 3M's

Electronic Products Division plant in Columbia, Missouri. The unit was discharging into the sewer wastewater contaminated with a developer solvent (1,1,1- trichloroethane). Tighter regulations prompted the unit's work force to consider how they could recover the solvent. They installed a closed-loop decanter system that separates the solvent from the water; the solvents are then distilled and reused.

Water consumption and pollution can also be reduced through recycling programs. Volkswagen's Wolfsburg plant has cut water use by 40 percent since 1973, largely through achieving a water recycling rate of almost 90 percent. Textiles companies can lower water use by installing countercurrent washing: the least contaminated water from the final wash is reused for the next-to-last wash and so on, until the water reaches the first wash stage, where it is discharged from the system and treated. Furthermore, the use of hot rather than cold water in some cases halved water consumption in the textile industry.[23]

Sometimes more fundamental changes, such as moving from a chemical to a mechanical process, can help prevent pollution. Vulcan Automotive Equipment, a small Canadian auto-engine remanufacturer, cut raw material, labor, and waste management costs while improving product quality by replacing its traditional inorganic caustic cleanser with a high-velocity aluminum shot system. Caustic soda had been used to clean caked oil and grime from old engines, creating health hazards for workers and a large amount of waste sludge. In its place, Vulcan introduced a two-step system in which the metal parts are initially baked to remove volatile oils and grease, and then sprayed with a high-velocity stream of aluminum shot to remove remaining dirt and rust.[24]

Resource Recovery

Pollution emissions can also be reduced by keeping the polluting agents within the production system, and reusing them in the same or other processes. Some industries have already established complex "industrial ecosystems" whereby the waste from one process becomes the feedstock for another.[25] Many large-scale petrochemical operations have extensive recycling circuits, which are used to return materials such as solvents or catalysts to the beginning of the process.

Car companies now regularly recycle production waste. General Motors separates scrap polyvinyl chloride (PVC) plastic by color. Then it grinds, melts, and reuses it, along with new PVC.[26] At Volkswagen, waste

> *"It is one of the tasks of our age to ensure that not only are the products of industry accepted and enjoyed, but that the consequences of industrial production are kept under control. If we think of the future—a central point of the obligation to rising generations—we must adopt the cyclical processes on which the whole of Nature is based."*
>
> Carl H. Hahn
> Vorsitzender des Vorstandes
> Volkswagen AG

thermosetting plastics used in the production line are collected and returned to the supplier. They are then reconditioned and mixed with 20-percent new material, and reused as soundproofing material in a Volkswagen model.

Materials that are waste for one industry can be useful inputs in another. At Du Pont, nylon production left behind 3,600 tons of hexamethyleneimine. Researchers discovered a ready market for this in the pharmaceuticals and coatings industries. Demand now exceeds supply, and in 1989 Du Pont found it profitable to make the "waste" on purpose.[27]

Waste exchange systems have been established in a number of countries to overcome the information gap between waste producers and potential customers. Organic wastes are now often being considered for use as agricultural fertilizers. In China, the organic wastes of thousands of small straw pulp mills are used for this purpose; in Denmark, the Novo-Nordisk biotechnology company has decided to convert the nitrogen wastes from its Kalundborg plant into fertilizer, which it distributes free to farmers.

This is only one of the increasingly complex set of material flows between Novo, the Asnaes coal-fired power station, a Statoil refinery, and a Gyproc plasterboard factory. For example, Novo buys "waste" steam from the power station, while the power station buys "waste" cooling water from the refinery. Known locally as "industrial symbiosis," this system has evolved over time in accordance with the twin needs of commercial viability and environmental quality. In the future, Asnaes plans to buy the flare gas from the oil refinery, while Gyproc is studying the possibility of using the gypsum produced by the desulfurization scrubber that will be installed in the power station in 1993.

In India, more than 30,000 metric tons of solid waste known as willow dust are produced by the textile industry each year. This is now being used to produce biogas, cutting energy consumption within the industry.[28] In Germany, BASF reports that it already produces 71.5 percent of the steam it needs with reclaimed heat from chemical processes or by burning residues.[29]

What separates a "waste" from a "raw material" is economic usefulness, and all these cases demonstrate how research and imagination can turn wastes into resources.

What Is Preventing Pollution Prevention?

Despite the widespread espousal of a waste management hierarchy that makes prevention the top priority, most government funds and regulatory efforts are still instead geared toward control.

In the United States, "over 99 percent of Federal and State environmental spending is devoted to controlling pollution after waste is generated. Less than one percent is spent to reduce the generation of waste," according to a U.S. government report.[30] For the OECD as a whole, scarcely more than 20 percent of pollution control investments are made in clean technologies.[31]

Three broad types of obstacles to the spread of pollution prevention within industry can be identified: economic, information, and management attitudes.

First, companies must be convinced that the introduction of a new, cleaner technology will really cut production costs: "For the firm, it is not only a matter of comparing two treatment costs (add-on and clean technology) but two production methods," according to an OECD report.[32] Going to the source of a pollution or resource use problem challenges existing ways of doing business, and can thus be seen by industry as more risky.

Although substantial gains can be achieved through improved efficiency and better housekeeping, there comes a point where significant technological change and investment are required. Smaller companies and those in developing countries are often not in a position to make such investments. Capital is scarce in all industries, and investments in pollution prevention must compete with other seemingly more profitable projects for funding. Companies may have invested large sums in existing capital equipment that need to be written off; end-of-pipe control systems thus appear more attractive.

The second obstacle is a lack of information. Practical data about pollution prevention options may be unavailable, while much information that is available—and promoted by environmental technology companies—stresses end-of-pipe solutions. Companies often need local examples from their own industrial sector that demonstrate both the benefits and the feasibility of pollution prevention. Within companies, information on environmental impacts is often poor, making it difficult to assess prevention projects.

The third and most important obstacle is management attitude. Many managers believe that environmental protection inevitably costs money, that it is a peripheral issue and a diversion from basic corporate goals. This attitude has been supported by the use of end-of-pipe controls that are unproductive and simply add costs to the business.

Even companies that have adopted environmental protection as a corporate goal often miss the full potential of pollution prevention by compartmentalizing responsibility: the knowledge that "someone else is responsible" is a barrier to understanding waste management problems and opportunities. George Moellenkamp at DRT International, one of Europe's leading environmental auditors, has found that "in most companies, organisational issues account for more than 90 percent of environmental problems."[33]

Success with pollution prevention ultimately comes down to desire: "Source reduction is more than an economic incentive or a compliance requirement. It is a priority for environmental stewardship against which we must continuously measure our performance," according to Paul R. Wilkinson of Du Pont.[34]

Developing-country industries may be less able to afford investments in pollution prevention, but they are also least able to afford the loss of resources and efficiency represented by traditional patterns of industrial development. During the 1980s, for example, a large fiber plant in South India was on the point of closing down because deforestation was threatening supplies of wood pulp. Concerned about deforestation, the state government had raised the raw wood price, threatening the company's competitive position. Management took a hard look at efficiency, measuring resource flows and carrying out more than 200 projects to improve efficiency between 1980 and 1989. Energy and chemicals consumption was cut by 50 percent, and as a by-product pollution control costs fell by more than 40 percent.[35]

Developing countries also cannot afford the costs of waste management, which within industrial nations are regarded as a heavy financial burden on the economy, according to the OECD.[36] However, cleaning up hazardous waste sites is even more expensive, and is expected to cost hundreds of billions of dollars in the United States alone.

By avoiding such burdens through pollution prevention, developing countries could increase their international competitiveness while improving environmental quality. Whether they can "leapfrog" the industrial world in this way depends greatly on the quality and effectiveness of the technology that is transferred from North to South. (See chapter 8.)

In developing countries, collaborative efforts between governments, research institutes, and industry can stretch resources. Collective action is particularly important for small and medium-sized enterprises, enabling them to spread the costs of environmental improvement through shared wastewater treatment plants, for instance. In Africa, the Kenya Industrial Research and Development Institute (KIRDI) has established an extensive program of assistance for local industry. In particular, KIRDI's Leather Development Centre has helped diffuse environmental "best practice" among the country's tanneries.[37]

Cleaner Products Through Life-Cycle Stewardship

As companies become better at preventing pollution and husbanding resources, attention is shifting from problems caused by production to those caused by the product itself. The two most prominent global environmental issues of the late 1980s—climate change and ozone depletion—highlighted the scale of damage caused by the uncoordinated consumption habits of billions of individuals.

In a sense, the dividing line between process and product is artificial; many products are inputs for other processes (such as plastics for auto components, and steel for auto chassis). Furthermore, the industrial ecosystem of the manufacturing process is only one part of a much wider ecosystem, which contains all flows of materials and wastes produced through the production, consumption, and disposal of goods and services.

But the present design of the industrial ecosystem is flawed: rather than acting according to the circular principles of natural ecosystems, the flow of goods and services is essentially linear. Products are produced, purchased, used, and dumped, with little regard for environmental efficiency or impact.

Recycling materials after products have reached the end of their useful life can bring considerable savings, as it avoids the stages of extraction and processing, which are energy- and pollution-intensive. For example, 47-74 percent less energy is used to produce steel from scrap material instead of from iron ore, while air and water pollution are cut by more than 75 percent.[38]

Corporate environmental responsibility no longer ends at the factory gate; it extends from cradle to grave. Ultimately, this means manufacturing only products that can be used within an environmental management system, which minimizes environmental impacts and maximizes environmental efficiency. This will require the construction of new commercial infrastructures and new relationships between producers, consumers, and governments.

While the notion of life-cycle environmental management came to prominence as recently as the late 1980s, corporate and government attention to the environmental impacts of products has existed since the beginning of environmental concern. Some products, such as petrochemicals, have had to be closely regulated to minimize inherent risks to health and the environment. Others, such as cars, have faced classic end-of-pipe treatment with the mandating of catalytic converters to reduce nitrogen oxides and hydrocarbon exhaust emissions. The State of California has begun the process of requiring the sale of "zero emission vehicles" that prevent pollution by using alternative fuels.

But many companies have found that the need to maintain product quality can inhibit pollution prevention. For example, 3M wanted to move from a solvent-based to a water-based process at a plant that manufactured audio tape. Yet despite extensive research, the water-based tape was found to be inferior, and the project was scrapped for quality reasons alone.[39]

Product reformulation has been seen largely as a way of reducing waste and pollution for 3M production sites. But companies are realizing

Table 7.1
Corporate Options for Product Improvement[40]

---

* Eliminate or replace product
* Eliminate or reduce harmful ingredients
* Substitute environmentally preferred materials or processes
* Decrease weight or reduce volume
* Produce concentrated product
* Produce in bulk
* Combine the functions of more than one product
* Produce fewer models or styles
* Redesign for more efficient use
* Increase product life span
* Reduce wasteful packaging
* Improve repairability
* Redesign for consumer reuse
* Remanufacture the product

---

that to capture and retain customers in an increasingly environmentally conscious marketplace, the products themselves have to be cleaner. Now, according to 3M's Allen Aspengren, the challenging question is, "What waste are we creating for our customers?"

Thus 3M is having to reexamine some of its products. Nevertheless, Aspengren believes that "environmental constraints have helped us to focus our efforts and develop better and more-efficient products." For example, the need to eliminate the use of PVC in products destined for the computer industry has resulted in a product reformulation that is "less expensive, better for the environment and better quality."

Companies wanting to improve their products' environmental performance can choose from a broad range of options. (See table 7.1.) These pose new challenges for product designers who now must consider such issues as recyclability, durability, and repairability in choosing technologies and materials. Clearly, the most radical solution is to remove a product from the market. In fact, some chemical companies have voluntarily withdrawn products they believed posed too great a risk to the environment. Ciba-Geigy, for example, has taken about 40 dyes off the market because it was unable to change the production process sufficiently to make it both economically and environmentally efficient.

Just as companies are under increasing pressure from regulators to reduce or eliminate certain environmentally harmful materials in the production process, there have also been moves to use more environmentally compatible materials in the end product. In Germany, the threat of legal action against the alleged health side effects of its cleaning products prompted Werner & Mertz to reformulate its products and launch a new range using the logo of a smiling green frog, replacing its earlier one of a red frog. And the HENKEL group spent almost 20 years developing an alternative to phosphates in washing powders. Now all its products sold in Germany are phosphate-free, and it has launched phosphate-free brands in other European countries.[41]

Managing a product life cycle for minimal environmental impacts poses tough conceptual and operational challenges for business. Each step in the life of a product has implications for the environment, often giving rise to a number of issues. Developing adequate tools with which to assess these environmental impacts is the starting point. Business, research institutes, and governments are working to develop life-cycle analyses (LCAs) or "eco-balances" to evaluate the cradle-to-grave implications of different product options.

The first LCAs in the late 1960s and early 1970s tended to focus on the comparative energy consumption of different materials, particularly for packaging. But LCAs go far beyond studies of energy balances. They are also used to evaluate resource requirements and environmental impacts. Each LCA has three parts: first, an inventory of energy, resource use, and emissions during each step of the product's life; second, an assessment of the impact of these components; and third, an action plan for improving the product's environmental performance. As LCA is still a relatively new concept, most to date have focussed on the inventory stage only.

It is important that life-cycle analyses are seen as a tool and not as a panacea for resolving the complex environmental issues involved with every product. Their purpose is to stimulate action and improvement. Procter & Gamble is one of the leaders in life-cycle analysis and has used it to illuminate a number of controversial issues, such as the relative merits of disposable diapers (which it produces) and competing cloth diapers. The company has also used LCAs to identify ways of reducing packaging waste. This has led to packaging innovations, such as maximizing the use of recycled materials, selling detergents in a concentrated form, and introducing refillable containers.[42]

Life-cycle analysis implies life-cycle responsibility. A combination of increasing external pressures and growing internal commitment has made some leading companies ensure that their products are made, used, and disposed of in the most environmentally compatible ways.

Some chemical companies, for instance, have product stewardship schemes that provide information and advice to their customers on how best to use their products. Declining landfill space for scrapped cars has prompted several car manufacturers in Germany and France to embark on total recycling schemes, aiming to reuse as much of the discarded vehicles as possible in the production of new cars. One company that has now pledged to take back used cars is Volkswagen.[43]

Retail: The Environmental Go-Between

In the words of Antonia Ax:son Johnson, chairman of Axel Johnson AB in Sweden, "A store can become an exhibition hall for conveying the concept of sustainable development at a very down-to-earth level." As gatekeepers between manufacturers and consumers, retailers have many opportunities to exert pressure in favor of sustainable development.

The emergence of green consumerism in North America and Europe in the late 1980s highlighted the role that the retail sector can play. Changes in technology, such as the use of bar-code scanning devices, have resulted in a shift in the balance of knowledge away from the manufacturer. The retailer now has the most detailed information on the movements of goods and knows what type of consumer buys what products and how often.

Not only have consumers become environmentally concerned—children encourage their parents and spouses encourage one another to be more environmentally sensitive. Retailers act as go-betweens, communicating the new demands of customers upstream to their suppliers and delivering new products and services downstream to those customers.

In addition to picking up the trends that lie behind millions of separate consumer decisions each day, the retailer can also act as an educator, providing data and analysis to help the customer make better informed choices. In Japan, companies complain that the spread of environmental products, particularly using recycled materials, is still held back by

> *"The change in corporate strategy requires the establishment of an environmental management system that includes grasping the connection between business activity and its environmental impacts from a long-term perspective, reassessing corporate philosophy, activities in research and development, production and sales, and finally evaluating the progress."*
>
> Toshiaki Yamaguchi
> President
> Tosoh Corporation

consumer apathy or hostility. This will only be overcome with more information and education.

Responding to the rising wave of public concern about the state of the environment, many retailers, particularly the large supermarket chains, have used their purchasing power to stimulate manufacturers to change their products. Some supermarkets have refused to stock products they consider environmentally unacceptable.

For example, as part of a long-running environmental program, the Tengelmann group in Germany stopped selling CFC-powered aerosols in 1988, and since 1989 the only batteries it has sold are mercury-free.[44] In 1991, the British "do-it-yourself" group B&Q decided to stop selling peat—used for gardening—that had been extracted from conservation sites. Many supermarkets have launched their own brands of environmentally friendly products to capture the emergence of a new niche in an increasingly saturated market, with slogans such as Nature's Choice (Loblaws, Canada), Ecologic (Safeway, United Kingdom), and Green Pledge (Prisunic, France).

At the U.S. mail-order company Smith & Hawken, concern about the use of tropical hardwoods in its furniture range led company president Paul Hawken to initiate a year-long inquiry into the situation. As a result, Smith & Hawken switched suppliers to ensure it was receiving timber from sustainably managed forests, and funded an industrywide certification program.[45]

One retailer that has taken an active stance in promoting sustainable development is the Swiss Migros cooperative.[46] Migros and others have shown that retailers can exert considerable influence at all phases of their involvement in the life cycle of products they sell, including the production, distribution, point of sale, and after-sales stages.

But the keys to the success of retail-driven sustainable consumerism are credibility and clarity. Consumers have become more demanding with retailer and manufacturer claims; slogans and declarations of intent are no longer enough. One way companies have sought to overcome the potential credibility gap is to obtain a stamp of approval from environmental organizations. The World Wide Fund for Nature (WWF) has long engaged with business in fundraising activities that allow companies to carry WWF's famous panda logos on their products. Such joint ventures are not always trouble-free: when Loblaws signed an agreement for Pollution Probe in Canada to endorse a range of "green" products in return for royalties, the director of Pollution Probe was forced to resign because of staff discontent with the deal.

Business recognition of the need for clarity in advertising has led to a decline in the use of ambiguous claims such as "environment friendly," and to government or industry moves to regulate the unsubstantiated use of advertising claims such as "recyclable" or "biodegradable."

Official eco-labelling schemes, such as Germany's Blue Angel, Canada's Environmental Choice, and Japan's Eco-mark, can both assist consumers in their choices and act as a stimulus to companies to design better products. By the end of 1991, nine member countries of the OECD had introduced labelling schemes; by 1992, "there could be as many as 22 OECD countries offering products with environmental labels on their market shelves," according to an OECD report.[47] In December 1991 the European Community adopted legislation establishing a Community-wide labelling program.[48] Nevertheless, these schemes would not claim to be fully comprehensive, often choosing one environmental feature out of many as the basis for awarding the label.

The goal is to make the manufacture, use, and disposal of products more compatible with sustainable development. It is not enough to restrict environmental concerns to a particular product range; all products need to be screened with environmental criteria in mind.

• *Production*: Wholesalers and retailers need to insist on open and continuous information from manufacturers on the production processes with respect to energy consumption, use of raw materials, and environmental pollution. Some retailers are already buying mainly from manufacturers who demonstrate a commitment to sustainable development. New supplier codes are being drawn up, setting out process and product requirements to help manufacturers minimize impacts throughout the product life cycle.

• *Distribution*: Retailers have a responsibility to ensure that goods are distributed efficiently. That often means using rail versus truck transportation for long distances, and specifying city diesel or non-fossil-fuel-powered trucks instead of conventional ones for the remainder of the journey. Retailers can also aim to avoid empty or nearly empty trucks or bulk containers on return trips.

• *Point of sale*: Clearly the store itself should be both environmentally attractive and sustainable in its resource use. As Bill Lindsey, environmental project engineer at the VeryFine chain in the United States, says, "We are thinking of habitat maintenance. And not just the habitats of critters that used to live where we build, but also the working habitat of our employees."[49] A first priority is to conserve energy. At ICA in Sweden, heat exchangers and low-energy lighting have been installed to increase efficiency. Dagab, a Swedish wholesaler, is using waste heat from freezers to heat its warehouses. Pick'n Pay, a retailer in South Africa, is saving electricity by incorporating natural light into the design of new stores.[50] Retailers have also taken a lead in replacing ozone-depleting refrigerants.

Clear and concise information on the environmental impacts of different products should be an integral part of display and sales activities, and some retailers have introduced environmental education kiosks.

Some stores also try to reflect a product's full environmental costs in the prices they charge. A number have started to charge for disposable plastic bags, or have dropped them altogether. At Seiyu chain stores in Japan, consumers are given a coupon for not using plastic bags; 20 coupons can be exchanged for 100 yen (80¢).[51]

Pricing is one of the most effective ways of changing consumer behavior. Today, many environmentally friendly products are more expensive. With pricing that better reflects environmental impacts (as called for in chapter 2), the prices of environmentally sound products would be lower than "dirtier" competing products.

• *Post-sales*: The growth in packaging is one of the most notable signs of the consumer society. As Antonia Ax:son Johnson of Axel Johnson says, "certain fast-moving consumer goods packagings are show-pieces of wasteful use of resources—toothpaste, for example. Three or four layers of packaging are used, without adding value."

Retailers are under increasing pressure from consumers and governments to help minimize packaging waste. In Germany, 80 percent of plastics and paper and 90 percent of glass, tin, and aluminum must be sorted and recycled by July 1995. If these quotas are not met, the government plans to impose a mandatory deposit of up to 50 pfennigs (about 30¢) per item from January 1993 on nearly all types of packaging. Consumers can also return packaging to retailers, who in turn can pass it on to the manufacturers, and so on up the chain to the packaging producer.[52]

The Challenges Ahead

Pollution prevention is best seen as a process of continuous improvement. But this does not mean that companies should not aim their sights high. Some U.S. companies such as Monsanto and General Dynamics are now setting "zero pollution" targets for certain substances, mirroring the "zero defects" pledge in the total quality management field. Monsanto chief executive officer Richard Mahoney believes that although a goal of zero emissions is scientifically impossible to achieve, it is "the only goal which will keep us stretching for ever greater improvement." At the company's Fisher Controls International plant at Marshalltown, Iowa, a zero discharge goal was set in 1988. Since then, water consumption and nonhazardous waste have been cut in half, and hazardous waste has been reduced by 90 percent. Plant managers in Iowa expect to eliminate the generation of all hazardous wastes by 1992.

As defined by the World Commission on Environment and Development, sustainable development stresses the importance of meeting human needs, placing an emphasis on satisfying the fundamental requirements of the poor for food, shelter, clothes, warmth, and education. Business leaders are beginning to appreciate that—in the words of Du Pont chairman of the board Edgar Woolard—"in a sustainable economy, world needs rather than existing product portfolios will determine which businesses we enter and which we leave."

For example, Du Pont is adapting to the needs of the developing world a flexible plastic pouch used to package milk in North America. In India, cooking oils have traditionally been ladled from large drums into the consumer's own container, carrying a risk of possible adulteration and contamination. Du Pont is working to modify the milk pouch so it can contain edible oils, enabling the consumer to buy the amount needed without threatening the product's quality and the consumer's health.

In the developing world itself, there appears to be increasing signs that environmental action can help generate employment, and thus satisfy one of the most basic human needs—for meaningful work. A report from the African Centre for Technology Studies has concluded that "promoting environmental conservation can be a major source of industrial activity and employment generation."[53]

The conceptual and technological foundations for achieving cleaner processes and products have already been laid. There is considerable and growing experience of pollution prevention, and companies are beginning to forge ahead with product life-cycle responsibility. But sustainable development means more than this, and in the years ahead business will be challenged to achieve zero pollution emissions from production plants and to redirect product development to meet basic needs, including those of the poor.

8

Technology
Cooperation

"It is the business community that will undertake the principal task of actually delivering solutions."

Michael Heseltine
Environment Secretary, United Kingdom

The requirement for clean, equitable economic growth everywhere—but particularly in the developing world—remains the single greatest problem within the larger challenge of sustainable development. Technology transfer—the movement of the technology required for economic development from where it exists to where it is needed—has long been a contentious issue in discussions between industrial and developing countries. It has usually been a clouded issue, for "technology transfer" is often seen by both sides as a euphemism for the transfer of capital.

The issue has reemerged with new vigor and new complexity on the sustainable development agenda. It was one of the main talking points in the preparatory meetings for the 1992 U.N. Conference on Environment and Development, and many reports have been produced. We do not intend to rehearse and analyze that debate. Instead we would suggest that technology transfer as a concept does not adequately capture the nature of the challenge posed by sustainable development. We suggest the term technology cooperation, which entails a broader range of objectives and is sharply focussed on business development. In our view, technology cooperation should put particular emphasis on building up the infrastructure, wealth-generating capacity, and competitiveness of a country.

Technology cooperation concentrates on developing human resources by extending a country's ability to absorb, generate, and apply knowledge. In developing countries, it works to enhance use of technology,

promote innovation, and foster entrepreneurship. Technology cooperation works best through business-to-business long-term partnerships
that ensure that both parties remain committed to the continued success
of the project.

The New Urgency

Multinational corporations (MNCs) have traditionally been among the
primary agents of technology transfer. They account for about a quarter
or more of manufacturing in Latin American countries such as Brazil,
Mexico, and Argentina, and also in Singapore and Malaysia. They may
be responsible for a third or more of the manufacturing exports of these
nations. Affiliates of U.S. MNCs in the mid-1980s accounted for 40
percent of all exports of machinery and 20 percent of all chemical exports
from Latin America. In Asia, the equivalent share was lower, about 7
percent on average, but was up to 20 percent in specific industry sectors
and countries.[1]

But recently concern for technology for development has been super-
seded by concern for technology for sustainable development. The
industrial world has taken this new emphasis seriously because of the
global impacts of environmental damage. Chlorofluorocarbons (CFCs)
released in China, for example, degrade the protective ozone layer over
North America; carbon emissions in Latin America may play a part in
changing Europe's climate and raising sea levels around its coasts. The
public debate over whether MNCs should have the same standards in
every country has also fuelled new discussions of this issue.

The mutual interest in technology transfer has been reflected in international environmental treaties. The 1979 convention on long-range
transboundary air pollution contained only a vague call for technological
transfer. Subsequent treaties and protocols have become steadily more
specific in mandating technology exchange, albeit in words rather than
deeds. The 1989 Basel convention on hazardous waste obliges signatories
to use appropriate means to cooperate in order "to assist developing
countries in the implementation" of the treaty, and "to cooperate in
developing the technical capacity among parties."[2]

The 1989 Montreal protocol on the phased replacement of CFCs
stressed the importance of ensuring access by developing countries to
"environmentally safe alternative substances and technology."[3] Its 1990

London amendment not only required signatories to transfer the best technologies on fair and favorable terms, it created a multilateral fund to help developing countries with the extra cost of meeting emission standards. It remains to be seen whether this sets a precedent for other agreements.

Lessons Forgotten

Much more has been written about technology transferred through official development assistance—a term that refers both to aid money channelled from government to government and through multilateral organizations—than about that moved by the private sector.

A key lesson of that history is that little technology can be moved successfully from one culture to another by a central organization, such as a development agency, because the "software" of a technology is at least as important as the hardware. Software here refers not only to the know-how, operating, and maintenance skills associated with the technology, but also to adaptations appropriate to the cultural context and previous experience of the receiving organization and the society that is going to use it. The software must match their level of education and skills, and enable users to operate the technology efficiently and cost-effectively. Software also includes the communications and other training tools to be provided by the technology's originator.

Aid agencies have difficulties with their side of the transfer because of the many constituencies, often conflicting, they must satisfy. For example, the first concern of any aid official must be the policies of his or her own country, rather than the needs of the country receiving the equipment. If a large share of the goods given in aid must be purchased from industries in the donor country, then the official must also worry about the needs of the domestic industry. Since official aid is mostly a government-to-government operation, the next concern is for the policies of the recipient government. Last, and too often least, the aid official is concerned about the actual needs and capabilities of the group receiving the equipment.

Given the complexity of such an agenda, it is no wonder that technology transferred strictly on a governmental basis is seldom appropriate to the needs and conditions of the people it is meant to serve. And it is not surprising that much of the developing world is littered with agricultural, medical, processing, and industrial equipment that is idle for a lack of spare parts, fuel, trained maintenance workers, supplies of raw materials, or a market for products.

In many aid-giving industrial countries, large contracting industries have profited handsomely from involvement in big aid-funded construction projects: ports, power stations, dams for water supply and irrigation, and so on. In far too many cases, such aid has focussed mainly on capital-intensive hardware that has prestige appeal for officials of the recipient government. A contractor procures the necessary technology under license from an operator in the industrial country, but the specialized management, maintenance, environmental, and operating know-how of this supplier—essential for the safe and efficient operation of the facility—is not included in the project. When the ribbons have been cut, both politicians and the contractor leave satisfied, but the facility often fails to operate at an acceptable level of productivity, and has adverse impacts on the environment and the surrounding population, whose traditional livelihoods may have been destroyed.

In contrast, technology cooperation suggests that time is needed for both sides to communicate, educate one another, and adjust the nature of the technology and the nature of the transfer. And it ideally implies longer-term commitment to the success of a business venture.

What Works: Commerce and Competition

Technology cooperation is likely to be most successful when it happens within a commercial setting—that is, when it involves commercially beneficial cooperation between two companies. This cuts through the conflicting motives of the aid agenda. Both the provider and the recipient companies will have clear, self-interested motives to make the deal succeed. The recipient will have tried to see that the software is right so that the investment provides long-term returns. But the providing company will also have taken some trouble to respect the needs of its customer, if only in the interests of future business and developing a market in the region.

A few participants in the technology transfer debate have taken the naive view that corporations can somehow move technology on a concessional basis. This is impossible. Business enterprises exist to generate wealth by adding value. Their level of return on investment is

> *"There are obvious opportunities for those enterprises adopting the principles of sustainable development early. There are markets for environmentally sound products, be it in regard to their production processes or characteristics (for example, biodegradable products). New business opportunities will emerge in the field of business counselling on environmental matters, design and implementation of production processes, and methods to improve the efficiency of the production processes to create less waste and recycle as much as possible."*
>
> Roberto de Andraca
> Chairman of the Board
> CAP S.A.

a measure of efficiency. Although many companies are owned or controlled by the public sector and are instruments of public policy, they must also meet these goals if they are to provide benefit on a competitive and efficient basis. Business and industry can make technology available only on competitive, commercial terms.

Once the principle of commercial cooperation has been established, companies are free to explore the many possibilities available within commercial realities. They may be willing to cooperate on technology at less than market rates in the interest of developing a market in a region, or of "corporate citizenship." But such decisions are solely those of the enterprises.

There will also be a growing number of situations in which both governments and companies see benefit in certain new technologies being available in a developing country although no business enterprise there is able to afford it. Funding such cooperation becomes a political issue.

An aid agency may then become involved in the transaction. In fact there may be—and often are—four chief parties in such an arrangement: the donor and recipient governments, and companies in the donor and the recipient countries. The technology will still tend to be used more efficiently if a firm in the private sector is using it. This fact is understood increasingly in the developing world, where a growing number of governments are privatizing bodies that provide water, power, and agricultural inputs.

Two examples of technology cooperation offer lessons here—one negative in terms of sustainable development and one positive. The first is the international arms trade. In the late 1980s, military spending by developing countries roughly totalled their spending on health and education—$170 billion a year, $38 billion of which was spent on importing arms, mostly from industrial countries.[4] It has been estimated that annual arms spending by developing countries is equivalent to 180 million person-years of income, based on the average income in those nations.[5] Thus when a need is perceived, money is found. Expensive efforts are also made to train people to use this military technology, to see that the equipment is efficiently absorbed. If governments begin to see environmental degradation as a security threat equal to that of armed opposition, sustainable development technology may command a higher share of the government's budget for purchasing and training.

The other, more positive, example is the technology cooperation behind the spread of Green Revolution hardware and software. For about two decades agricultural research laboratories around the world have been developing new crop varieties, chemicals, and farming techniques under the coordination of the Consultative Group on International Agricultural Research. The effort began with private funding from the Rockefeller and Ford foundations, and today there are typically more than 30 donors in a given year, including governments, U.N. agencies, and the World Bank.[6]

Private companies have helped develop chemicals and seeds commercially, and have sold them. This effort has greatly increased yields of corn in Mexico and the Americas, of wheat in Central America and India, and of rice throughout Asia. The education and training of users, who went on to become innovators, was always a large part of the package. Although concern has been expressed about the sustainability of some Green Revolution techniques, this work provides an excellent model of how various sectors can cooperate in long-term partnerships for technology cooperation.

The Changing Scene

Many recent developments suggest that technology cooperation for sustainable development is about to become a major focus of business, governments, and multilateral organizations. Clean technology has become available in industrial nations only relatively recently, largely because of false market signals. Yet as polluters are forced to pay for the damage caused by pollution, cleaner technologies will become the norm.

Chapter 7 described how rapidly clean technology and processes are being developed. Some of these are appearing first in developing nations, which suggests the importance of South-South technology cooperation. In fact, a growing number of examples can be cited of technologies moving from developing to industrial countries. For instance, when London's underground trains could no longer cope with heavy commuter traffic, an Indian computer company, CMC Ltd, was called in to overhaul the timetable system. CMC beat six London firms in a competitive bid for the job.[7]

Governments and businesses in developing countries have not been particularly keen on purchasing clean technology or seeking it through aid programs. Development assistance agencies have not encouraged its inclusion in their efforts partly because of this lack of enthusiasm in developing nations and partly because the concept of sustainable development based on international cooperation is only slowly finding its way into these agencies.

But this situation is changing. Governments of industrial countries are realizing that it may be cost-effective to protect their own environments by spending money to prevent pollution outside their borders. In 1990, the five Nordic countries established the Nordic Environment Finance Corporation to provide venture capital for environmentally sound investments in Central and Eastern Europe.[8] This is partly aid, but it also decreases the amounts of acid pollution from these areas, which damages the Nordic environment. Japan has similar initiatives concerned with coal-fired power stations in China. By the same token, industrial countries seeking to limit greenhouse gas emissions may find it far more cost-effective to curb them in poorer countries, where they can be cut at lower cost. All these trends will both support and require technology cooperation.

Also, multinationals are recognizing the need to be as clean abroad as at home; this will mean cooperation—if only with local employees—over new technologies. The Environmental Charter of the Japanese business group the Keidanren provides 10 guidelines for Japanese companies operating abroad. These include applying Japanese standards concerning the management of harmful substances, providing the local community with information on environmental measures, and cooperating in the promotion of the country's scientific and other environmental measures.[9]

Another trend supporting technology cooperation is that a growing number of developing countries are honoring and protecting patents and other intellectual property rights. This means it is safer for companies to bring in technology. Where they are respected, intellectual property rights are not the barrier to effective cooperation that many participants in the technology cooperation debate believe them to be. Clear rights to intellectual property are an essential condition for the global spread of technology. Many basic infrastructure technologies are in the public domain.

Some industries important to sustainable development are already widely and freely sharing their know-how. In the timber business, for instance, sources of supply are more important to success than planting and harvesting techniques, so companies are cooperating to exchange information. Mitsubishi Corporation has formed a new collaboration for technical cooperation in forestry among governments, academic institutions, and commercial enterprises.[10]

A final, and perhaps the most important, prerequisite for technology cooperation exists in a growing but still small number of developing countries. This is a governmental and social environment that encourages and supports the development of business and direct investment from abroad. We cover this necessity more thoroughly in chapter 10.

When considering where to invest, MNCs thoroughly study a wide range of variables. They examine the new market from the viewpoints of location, prospects for and expected location of market growth, customers' preferences, local production costs, and competition. They consider the form and stability of the government, the financial stability of the country, cultural compatibility, the ethics of the political leadership, taxation laws, possibilities of cooperation with other firms, and environmental standards. They assure themselves that the country has a trained work force, adequate educational standards, training facilities for crafts people, appropriate salary scales, and adequate housing and recreation facilities. Within regions, countries compete for such investments on the basis of these factors.

Just 10 countries in the developing world account for 75 percent of total foreign direct investments. These are—in order of size of investment—

Singapore, Brazil, Mexico, China, Hong Kong, Malaysia, Egypt, Argentina, Thailand, and Colombia.[11] Although a few of these are deficient in some of the following attributes, this list confirms the importance of macroeconomic and political stability, reliable legal and property systems, an educated work force, and an adequate physical infrastructure (roads, power, communications, and so on).

Today, less than a third of all countries possess the political, social, and economic framework necessary to conduct international technology cooperation activities on purely commercial terms. Discussions with multinational corporations belonging to the BCSD identified about 50 nations where such transactions were routinely successful. Interestingly, this list corresponds closely to the 53 countries identified by the U.N. Development Programme as having "high human development."[12]

A growing number of countries are attempting to join their ranks. In the meantime, many of the best examples of technology cooperation occur between West and East, as some East European countries are seen to have great capacity to absorb technology in the form of large numbers of trained personnel. At the beginning of 1991, there were a total of 14,640 joint ventures in Eastern and Central Europe, including the former Soviet Union. The estimated accumulated value of the foreign direct investments was about $4 billion.[13]

ABB, an electrotechnical equipment, energy, and transportation multinational with headquarters in Europe, initiated a joint venture in Poland in 1989 that converted a former state-owned operation into a privately owned company, ABB Zamech. A large program of retrofitting combustion and power generation technologies has led to both productivity gains and pollution abatement. ABB's quality management practices were transferred, together with environmental management practices, with extensive staff training. New licenses for the required state-of-the-art gas turbines will enable ABB Zamech to become a "center of excellence" for a specific gas turbine technology within the ABB group worldwide. Results of the venture for the first operating year gave a return on sales of 5.2 percent.[14]

ABB, Shell, and Nissan are among the multinational companies that intend to become "local firms" in many different countries, largely by establishing ventures that can evolve into autonomous national companies in each host location, managed by trained nationals and with a large measure of independence in financing, product development, research, marketing, and strategy. As a result, some MNCs will cease to be flag carriers of any particular country.

Long-Term Partnerships

Given the above favorable trends, multinational corporations should demonstrate—as many are already—a willingness to make long-term investments in countries at an early stage of development by entering into partnerships to build, own, and operate joint ventures using the technology of sustainable development.

Elements of such partnerships include a long-term commitment to business development, to the training of employees, to adapting, improving, and upgrading technologies, and to introducing new management systems. Key in this context is the adaptation and orientation of technology to the local needs of people and markets in developing countries. For example, technologies designed to meet conditions where labor is expensive and capital cheap will not be an effective solution where these conditions are reversed.

"Sustainable development calls for technology transactions that are economically efficient, commercially attractive, and at the same time environmentally acceptable," explained Avininder Singh, the chairman of the environment committee of the Confederation of Indian Industry. "Business potential can help with these concerns, and companies should now move toward leadership."

The simplest form of this international cooperation is technology transfer within multinationals, with their common culture and objectives. Even here, however, some formal agreement is needed to satisfy the authorities in participating countries. The cooperation involved must therefore extend to the national governments of both home and host countries. The next most straightforward form of cooperation is a purely commercial transaction between two independent parties with full understanding and equivalent negotiating skills.

This recommendation of long-term partnerships is hardly a new concept for successful multinationals, for whom technology cooperation is synonymous with business development, and is their principal means of expansion. Yet what makes profits for multinationals also transfers competitive advantage to developing countries. This was noted in a BCSD Workshop on Technology Cooperation in Kuala Lumpur, Malay-

> *"Business should not be confined to the back seat. Together with government, business should take the lead in promoting sustainable development."*
>
> Anand Panyarachun
> Prime Minister of Thailand
> Former Chairman of Saha-Union Corp. Ltd.

sia, by Darwin Wika, manager of environment, health, and safety for Du Pont Asia: "The competitive advantage of an enterprise is based on technological innovation. Transferring technology means transferring competitive advantage. It is therefore a sensitive issue. Companies want expansion of business through long-term partnerships, not loss of business by the selling of their technologies."

At the other end of the business spectrum, small and medium-sized enterprises (SMEs) play an extremely important role in economic development and thus potentially an important role in sustainable development. (See also chapter 10.) But they face additional barriers in participating in technology cooperation. The largest is a lack of capital. An almost equally formidable obstacle, however, is access to information and to the skills required for assessing alternatives and negotiating arrangements. The most important people in technology cooperation are the technology receivers, because the speed and success of the transaction depends on their technical skills and management abilities. If the cooperation is between independent parties and is to succeed rapidly, the SME involved needs a technological entrepreneur—a leader with both technical and management ability.

Few SMEs have such a figure, due to a lack both of successful role models and mentors and of appropriate training and opportunities to gain experience. With encouragement and training, their numbers will grow. This is very important for sustainable development, because technology-intensive SMEs in industrial countries are vital sources of new technologies and product innovation, as well as of new employment.

Speaking of developing-country SMEs at a BCSD workshop in India, K.P. Nyati, director of India's National Productivity Council, said: "They are not aware of different technologies available today, and do not even think along the lines of intensity of noise, pollution level, energy consumption, etc. These entrepreneurs are guided by overall profitability alone. The choices of technology are made depending on the economic aspect and thus more often than not they tend to select wasteful technologies." Nyati went on to say that the solution lies in giving SMEs information on manufacturing technologies at the community level, preferably through trade associations. He noted that the need for technologies appropriate and compatible with local skills and labor available could not be overstated.

SMEs should tackle their technology cooperation problems in three steps. First, organize for information exchange and mutual help in nongovernmental associations. Second, pressure government agencies to ease access to venture capital and to provide grants for the acquisition and assessment of information. And third, gain access to the seasoned expertise required for technical assessments and negotiations. Even the largest MNCs depend on the services of external consultants with special knowledge of examining new developments and opportunities. In an open competitive market, consulting firms with the requisite technical and other specialist skills will quickly make themselves available.

The Confederation of Indian Industry is a good example of a nongovernmental association that creates local networks of SMEs and larger companies to share information and management practices and to seek government help in gaining access to technology and fostering the conditions for entrepreneurship and innovation.

Associations representing a single industrial sector—such as the Canadian Chemical Producers' Association, which initiated the environmental management concept of Responsible Care within that industry—have also been quite successful in spreading technologies and new thinking.[15] At the municipal level, local technology networks of small and large firms, universities, and government agencies have helped meet the needs common to small and new businesses.

Another important source of information and technical assistance for SMEs is local subsidiaries of multinational corporations. "Technology cooperation will strengthen the ties between SMEs and the multinationals," explained Udo Uwakaneme, president of the Enabling Environment Forum, Lagos, Nigeria, at a BCSD Workshop. "SMEs are being influenced by multinationals in terms of safety, effluents, and quality standards. There is plenty of evidence to show that such standards are much higher when cooperating with multinationals rather than with nationals."

Multinationals also benefit. "We want to help local companies develop their production know-how, because that often can reduce our production costs," said Noboru Miura, managing director and general manager of the electronic engineering group of Nissan Motors, Japan. "It takes time to reach a world-class level of production, and local producers must understand and accept that." Selecting a technology appropriate to the ability and capacity of local partners is a key issue, Nippon Steel managers emphasized. "If a project starts at the right level of technology, it assures quicker progress toward higher levels," they pointed out.

Technology cooperation can take place between a multinational corporation and its smaller suppliers upstream or its smaller customers downstream. Many successful examples of both exist. Nissan's policy of localization includes detailed methods for transferring technology and management skills to different countries of manufacture. The sequence begins with local knockdown production—the final assembly of cars in kit form with less than 40 percent local content. As more responsibility is passed to local management, the content of parts manufactured there from Japanese designs and specifications increases. Indigenous research and development (R&D) follows, to enable production of parts modified to meet local conditions. As experience grows, decision making is also localized, and independent parts development takes place. The local content of the manufactured automobile passes 60 percent. Almost all responsibilities and decisions are then in the hands of the overseas operation.

This technology cooperation involves production equipment, engineering methods and process technology, production facilities, tools and dies, and suppliers' engineering methods and process technology. Each has accompanying specifications, manuals, and drawings. Residency and training programs at Nissan Motor Limited's head office in Tokyo deliver much of the software of the cooperation. An information network ensures standardization. Engineers and advisors from the head office participate directly in overseas operations and R&D. Such approaches are spreading rapidly.

Cooperation on Training

"The most important problem in technology transfer is linked to education, and consists of three parts: insufficient general educational levels, a lack of technology awareness among government officials, and a lack of technology training within the recipient organization," according to Kenjiro Kimura, managing director and general manager of corporate technology and planning for Kyocera Corporation, Japan.

"If we have a joint venture, of course we pay for training," said Makoto Yoshida, general manager, environmental control plant safety, Nippon Steel Corporation, "but in the case of overseas development assistance programs sponsored by the government, the funds often do not cover proper training or follow-up monitoring of technology transfer."

When Shell first began developing Nigerian oil fields, Nigerians tended to be in low-skilled, low-wage jobs. Shell began a training program meant to eventually replace expatriates in most key positions in the company, as well as to improve skills in all jobs. Its scholarship program sent employees overseas and to Nigerian universities. It established two local training centers. Nigerians now occupy 90 percent of all Shell jobs in that country, and meet the same technical standards required of Shell employees worldwide. The company has clearly benefitted, but the program also sent ripples throughout the Nigerian economy by providing a large group of technically trained personnel.[16]

Such individuals will be the innovators in producing technology for sustainable development appropriate to local conditions. Thus technology cooperation does not simply move technology, it also transfers the ability to innovate. Development is not only about improved living standards but also about a people's capacity to absorb, generate, and apply knowledge.

Innovative Aid

Except for the purposes of natural resource extraction, little foreign direct investment reaches countries near the bottom of the development ladder because the political and social risks of investment are generally too high. So extending the mechanism of long-term business partnerships to low- and middle-income countries requires security for the direct private investment that must be involved to ensure accountability and the success of the venture. This security starts with political stability, which provides a set of reliable rules and regulations.

Official development assistance can help governments collaborate with business and industry in delivering technology and management systems for basic infrastructure through long-term business partnerships. It can help lower the overall financial risks, and may indirectly provide strong leverage to promote political and social reforms and other structural adjustments.

Even with such added financial security, the residual risks for some projects will still remain too high in commercial terms. Recognizing this, the World Bank is testing a way to "provide comfort against sovereign risks" with the Hab River project in Pakistan by ensuring, for example, that loan payments and dividends will be repatriated and available in convertible currency. The commercial risks can then be financed by the business parties through normal commercial mechanisms. If this Bank experiment succeeds, this mechanism could be more broadly used. The combined result would be earlier industrialization, earlier development of markets, and hence earlier establishment of the business factors for national economic growth.[17]

We are not calling for aid to be diverted to business. And we realize that development assistance can also help technology cooperation by supporting institution building and education and training programs that increase countries' capacities to absorb and use new technologies. But it is obvious that sustainable development will require—along with international structural adjustments such as the removal of trade barriers to developing-country exports—an increase in total amounts of capital available for development. Some of this should facilitate technology cooperation.

Such an effort might include matching aid funds for foreign direct investment projects through joint ventures that build and operate facilities for sustainable development in low- and middle-income countries, guarantees against sovereign risks on such projects, grants to technological entrepreneurs in SMEs for access to information and consultant expertise on new business development and environmental protection, matching funds for private foreign direct investment in the form of venture capital for technology cooperation start-ups in developing countries, and accelerated approval and financing for pilot projects to demonstrate technology cooperation, accompanied by wide publicity about the results and methods used.

This is not all theoretical. In order to negotiate and sustain free trade with Canada and the United States, Mexico must improve its environmental performance to neutralize arguments that Mexican companies compete unfairly by not meeting strict international standards. A partnership has been formed between Canada-based Northern Telecom, the government of Mexico (the first to sign the Montreal protocol on ozone protection), and the U.S. Environmental Protection Agency (EPA) to eliminate ozone-depleting solvents from Mexican industry.[18]

Experts from an information-sharing organization, the Industry Cooperative for Ozone Layer Protection, held a series of workshops in Mexico to introduce alternative technologies. In cooperation with SEDUE (the Mexican environment ministry), EPA and Northern Telecom have created a model project for technology cooperation that takes advantage of the combined strengths of government and business. The ministry is providing intersectoral coordination and facilitating investment in modern technologies; EPA is providing support in the development of environmental control procedures; and Northern Telecom is sharing its experience in implementing processes and technologies, managing the workshops, and coordinating the input of experts from other companies.

The Chain of Technology Cooperation

Many businesses have begun developing long-term partnerships for technology cooperation such as those just described. Others will want to begin as soon as possible to take advantage of the new opportunities that a move toward sustainable development will produce.

But technology cooperation is a complex game with many players: businesses in industrial and developing nations; the governments of those countries; schools and all training bodies; and nongovernmental organizations, with pressure from the media and consumers to maintain the movement toward sustainable development. Technology cooperation for sustainable development is best seen as a chain of new linkages.

First, the growing competitive advantage from environmentally sound development and resource efficiency has been linked to a desire for the transfer of suitable technology. This desire must then be linked to the creation of long-term business partnerships for technology cooperation, so that these become the primary means of international business development. Such cooperative ventures must be tied to the technical and management training essential for safe, efficient, environmentally sound operations, which may take years.

Second, training must be linked to innovation, to produce not just skilled workers but innovators. Making this link involves the integration of personnel from foreign subsidiaries and joint ventures into relevant research and development activities, through training and personnel exchanges inside the company. The purpose is to join technology development to technology cooperation: recipients must be involved in technology development, and the development team must understand local conditions and needs so they can integrate them into the approaches to the technology.

Last, this innovation must be linked to the goal of sustainable development—and all the corporate management changes, new products, new processes, and new infrastructure that go with it. This completes the chain. It also shows how technology cooperation transfers competitive advantage internationally.

9

Sustainable Management
of Renewable Resources:
Agriculture and Forestry

"To care about the environment requires at least one square meal a day."

Richard Leakey
Director, Kenya Wildlife Service

Farming and forestry are central to sustainable development because of the high numbers of people working in both areas, the amounts of money generated, and the extensive, direct impacts that both have on renewable resources and the environment. Some 40 percent of global employment and 50 percent of world assets are associated with these two businesses.[1] Especially in developing countries, the health of farming and forestry and their resource bases have a major impact on nutrition, energy supply, employment, population growth, and rural migration. They are also linked to energy issues, as agriculture and forestry have energy needs and as forest biomass is a vital domestic energy source in many developing countries. And they are linked to issues of water protection and use.

National and international strategies for the use of such resources must be reoriented along the principles of more open, competitive markets and drastically reduced trade restrictions; more access to key production factors such as credit, land, and know-how; effective property rights, and land tenure and land use policies; and improved education, research, and management training, particularly in developing countries.

Business can best contribute to these efforts within the framework of the market economy, with a mix of economic instruments, clear performance-based regulations, and corporate stewardship based on setting international standards. Short-term, no-regrets policies must be supplemented with long-term research and development strategies aimed at economic growth with diminishing environmental impacts. Sustainable development in farming and forestry requires stronger cooperation between business, government, and local community groups.

Many similarities can be found between farming and forestry. But farming has tended to reside in the private sector down through the ages. State farms developed in this century have largely failed, and where farms have been returned to private ownership, yields have increased dramatically. Forests, on the other hand, have tended to have no effectively enforced ownership or various types of communal management responsibility systems involving tribal peoples. This has generally given way to government ownership, with about three quarters of the planet's forests now owned by governments.[2] The efficiency of government ownership and management of forests, at least for the production of timber, is now being questioned—not least by governments themselves. Initiatives for private business in forestry are being developed. Using sustainable private management, several countries have increased their forest inventory in the twentieth century.

Food and Agriculture

Food production rose at an average annual rate of 3.2 percent in developing countries and 2 percent in industrial countries between 1961 and 1985. Yields doubled to feed an additional 2 billion people over that period; about one third of the increase was the result of expanding harvested land areas, with two thirds due to higher crop yields from modern, intensive agriculture methods. Currently, about 15 million square kilometers of the earth's surface are used to grow crops.[3]

A larger proportion of the world's population enjoyed adequate nutrition during 1990 than ever before, yet there are 750 million malnourished people in the world today, and 75,000 people die each day (27 million per year) from malnutrition-related causes; most of these are children.[4]

It is often assumed that wealthier countries do more damage to the environment than poorer ones, but in the case of agriculture the opposite may be the case. Understanding the reasons for this is essential to achieving the goal of sustainable farming—a goal with three components: feeding a growing population, sustaining farm incomes to keep farmers in business, and protecting the earth's ability to continue providing food.

Industrial Countries

Many industrial countries enjoy low-cost food and surplus produce for export, and farmers there make a decent living, at least compared with most farmers in developing countries. Environmental harm from farming has been reduced in recent years, and environmentally sustainable farm practices appear to be within reach.

Many factors contributed to the rise in yields registered in these countries, such as fertilizers, herbicides, pesticides, mechanization, better crop varieties, and improved water management and soil conservation. According to a recent Hudson Institute study, without high-yield, intensive agriculture the world would probably have to triple its current cropland base to provide the same quantity of food. That would mean plowing an additional 26-28 million square kilometers—an area much larger than North America—resulting in a huge loss of forestland and wildlife habitat.[5]

U.S. farmers have controlled soil loss by terracing and watershed management and by increasing yields on good land, so as to reduce the need to plough up marginal, fragile lands. And they have adopted a range of new conservation tillage practices: reduced-till, ridge till, and no-till farming.[6] This has had the welcome side effect of reducing fuel use in the U.S. farm sector.[7] The 1990 U.S. Farm Bill added an integrated farm management option that encourages farmers to conserve soil through rotating their usual crops with small grains, grasses, and legumes.

Progress continues to be made in the control of the most toxic farm chemicals. The highly persistent organochlorine pesticides (like DDT, aldrin, and dieldrin), which once devastated wildlife in rural areas, have been banned from agricultural use in the United States and most other industrial countries for nearly two decades.

Farm chemical industries are now producing pesticides that degrade more quickly, that have more focussed effects, and that can be applied at lower doses. In terms of overall amounts, the volume of both insecticides and herbicides used on farms declined in the United States in the 1980s.[8] With techniques such as ultra-low volume and spot spraying, ingredients are better targeted. Biotechnology is expected to produce crop varieties naturally resistant to pests, thus permitting a further reduction in the most toxic chemical use in the years and decades ahead.

The methods developed in the United States to use agricultural chemicals more efficiently and safely are referred to as "best management practices." These techniques include such things as new crop rotation systems, the use of computers to guide chemical use, and integrated pest management (IPM). In IPM, farmers use biological controls and apply pesticides when pest levels reach economically damaging levels rather than on a certain calendar date; this controls pests, uses smaller amounts of pesticides, saves money, and causes less environmental damage. The practice of applying pesticides by the calendar is increasingly uncommon for cotton, tomatoes grown for canning, and other crops where IPM is having success. The U.S. government funds IPM research projects on more than 100 major and minor crops.[9]

Environmentalists who have encouraged such recent improvements in industrial-country agricultural practices are not satisfied with these results, nor should they be. A number of serious environmental problems continue to plague U.S. farms—including excessive water use and inadequate range management, mostly on Western public lands.

Fertilizer use rates in the European Community (EC) are about three times those of the United States. And in Japan the rate is about four and a half times as high.[10] Although differences in soil conditions, climate, and crop technology explain some of the variance, much of the difference is due to efforts to achieve domestic food security and support small farmers who cannot compete with lower-cost producers on larger farms in North America, Australia, Argentina, and other major food exporters. In the European Community, production stimulants built into the Common Agricultural Policy have led to a systematic overuse of nitrogen fertilizers, overproduction, and water pollution problems. Policies include economic incentives that guarantee prices for farm products and restrict competing imports. Removal of such support will force further rationalization of European agriculture.

To address the environmental and economic costs of intensive agriculture, changes in EC farm policy since 1980 have aimed at curbing the production of surplus commodities, displacing imports, reducing the cost of agricultural support programs, and lowering the environmental impact of agriculture. As in the United States, development of codes of good agricultural practice has been emphasized. Unfortunately, there has been little success to date in reducing surplus production or high support-program costs.

In 1987, the EC introduced an Environmentally Sensitive Areas scheme to pay farmers to manage their land in approved ways. In Baden-Württemberg, West Germany, nitrogen fertilizer use in areas of important freshwater supplies was limited to 20 percent below the optimal level. In 1988, Britain abolished all farm capital grants related to production, replacing them with grants for conservation. Current Dutch plans call for reducing nitrate and other emissions into rivers by 90 percent; a quota has been imposed on the numbers of Dutch pigs and other livestock, which is expected to be lowered even further in order to achieve the emissions goal.[11]

Given the other environmental costs of overly intensive agriculture, growing environmental awareness in Europe, and the press for an end to agricultural subsidies and trade barriers, European farmers are likely to cut back on their use of agricultural chemicals. Continued political pressures from consumers, heightened environmental awareness, and the continuing development of new technologies should help alleviate other problems as well.

The work of some environmental groups and their proposals for reforms and regulations remain important, and their political efforts have had an effect. These are linked to market pressures. Consumers of farm products are becoming environmentally conscious and health-conscious, and are pressing through the marketplace for more sustainable farming.

New scientific knowledge and technical innovation, much of it provided by business research and development, allow farmers to improve environmental protection and increase productivity at the same time. The relative wealth, education, and technical sophistication of farmers speeds the adoption of these new, more sustainable farming techniques.

Problems in industrial countries remain those of policies rather than technologies. Many of them are contradictory: they attempt to boost farm incomes while providing reasonably priced food to consumers; they try to increase productivity while stabilizing agricultural commodity prices; they support small farms, often in marginal farming areas, while trying to improve efficiency, conserve agricultural resources, and protect the environment. Price support programs associated with these policies create unintended problems elsewhere in the domestic and global economy.

Policies that subsidize some crops more than others and that attempt to control their production lead to farmers being more concerned with responding to commodity price subsidy programs than to market signals. Farmers respond by growing more of the more heavily supported crops and by using more chemicals, land, and energy. These policies are not cost-effective and create a huge burden for taxpayers, consumers, and the environment.

The Developing World

The same optimism about the impact of food production on the environment does not yet apply to the developing world, largely because the same preconditions for sustainability do not exist. In some developing countries, environmental problems in this sector, such as chemical pollution, have been evolving much as they did earlier in today's industrial countries. But in others, rural resource protection problems today bear little resemblance to those of the industrial world. This is because they are evolving in response to their own unique ecological and demographic dynamics. And in some degraded regions of rural Africa, modern machinery and farm chemicals are seldom to be seen.

In many poor countries, environmental destruction from farming is rapidly accelerating. Land degradation has emerged as the single most serious environmental problem in many developing countries.[12] Roughly half the irrigated cropland in the developing world now requires reclamation because of salinity or poor drainage.[13] Soil erosion and infertility are simultaneously degrading 30 percent of all rainfed cropland in Central America, 17 percent in Africa, 20 percent in Southwest Asia, and 36 percent in Southeast Asia.[14]

Much of the serious deforestation in developing countries occurs as farmers abandon degraded, previously productive fields to clear new land. Cutting down trees, however, usually speeds the degradation process, since tropical forest soils are rarely suited to continuous cultivation and intensive grazing. Another important reason for deforestation has until recently been fast-growing cattle production. Since 1950, world meat production has tripled. In Latin America, 20 million hectares of tropical forest have been changed to pasture since 1970.[15]

Farm chemicals are a growing environmental problem in parts of the developing world (especially downstream from irrigated rice cultivation

> *"I would like to emphasize that the most important natural resources are the human beings. Many countries in Europe and in Asia have very limited natural resources. Even so they are among the most developed countries in the world. Therefore I am not particularly impressed by physical natural resources, but by human resources."*
>
> Erling S. Lorentzen
> Chairman
> Aracruz Celulose S.A.

in Asia), and an acute human safety problem as well. The World Health Organization has estimated that as many as 1.5 million accidental pesticide poisonings occur every year, mostly in the developing world, where the training and equipment needed for worker protection are often absent.[16]

Despite progress in the North and growing problems in the South, a simple transfer of farm technology and techniques to developing countries is unlikely to be helpful, except in some with similar climates and more advanced economies. Rapidly developing Asian countries can probably safely and effectively adopt and adapt many such practices.

In countries where secure title to land does not exist, where cash incentives for farm resource protection are not yet affordable, where limited scientific and technical capacity constrains innovation, where human and capital resource deficits in the countryside hamper the adoption of new technologies, and where governments do not accept advice from environmental groups, environmental performance naturally lags.

In many areas, the problem of protecting rural resources has several further dimensions. The first is the inherent fragility of the natural resource base. Farming resources in many poor tropical countries are simply not as durable as in the industrial, temperate-zone countries, because of either inadequate rainfall or terrain or soil structure. If these fragile resources are not farmed with great care, they will quickly degrade. Traditionally, this was accomplished (albeit with little or no economic growth) through techniques such as limited cultivation, shifting or rotational cultivation, extensive tree or bush fallow systems, and tight social controls over fertility and migration.[17]

Such traditional environmental protection techniques have ceased to be effective in many places under the pressure of population growth— of people, livestock, or both. It is this combination of fragile resources with rapid population growth that makes the rural resource protection problem in many developing countries so difficult, and so unlikely to be resolved through the simple application of policies or techniques devised in industrial countries. Policies will have to be tailored to local conditions, by local groups.

Agriculture and Trade

International agricultural trade reveals some links between farming successes in the industrial world and failures in the developing world. Most food does not enter such trade; it is produced and consumed in the same country. Only 15-20 percent of world wheat and maize moves between nations, for example.[18] Yet for countries dependent on food imports, such trade can be crucial.

Through export subsidies, import restrictions, and internal agricultural subsidies, industrial countries have accelerated the falling trend in real prices for agricultural commodities traded. Tariffs on nonagricultural goods have been reduced significantly, while agricultural product tariffs have increased. This has made it more difficult for poor countries to develop their agriculture and to export food to pay for needed imports.

There is a conflict between the political objective of food self-sufficiency in industrial countries and the need of developing countries for an open trade in food products. A primary consequence of Northern agricultural policies is a shift in global production away from areas that may have comparative advantages in terms of labor costs, natural resources, and climate to areas that can and do support artificially high costs of production. These policies encourage overproduction in richer, more economically diverse countries with few farmers, and suppress production in those with less diversified or healthy economies, large farming populations, and often many hungry people.

Increased international food trade at market-determined prices could be a tremendous boon to developing countries. Chile has become a major exporter of fruits and vegetables, for instance, supplying off-season demand of consumers in North America and West Europe.[19] International trade reform, addressed in chapter 5, is essential for sustainable agriculture. Market forces can be used to encourage the growing of more food in developing countries in a more sustainable manner.

In the 1986 Uruguay Round of the General Agreement on Tariffs and Trade (GATT), the United States, Australia, New Zealand, Canada, and many developing countries called for the elimination of trade-distorting farm subsidies within a decade. Subsidies in world agriculture and trade barriers should be phased out as soon as possible.

Problems and Possibilities in Developing Countries

Arguably, none of the formulas for assisting developing countries that became popular during the 1980s—neither the structural adjustment approach of the World Bank nor the traditional technologies favored by many environmental groups—is fully adequate or appropriate to the task of improving agriculture in those nations.

The structural adjustment approach gives inadequate weight to needed public-sector investments in rural resource protection.[20] The traditional technologies approach overlooks the growth of rural populations, and the inability of larger populations to continue protecting rural resources with conventional methods. It will be essential to increase yields on resilient lands in order to reduce the cultivation burden on fragile soils. Environment and development groups operating in Southern rural areas should be working hard with local farmers to develop new cultivation techniques, to educate and train people, and to organize farmers to spread successful new technologies.

Governments will retain most of the responsibility for encouraging partnerships between farmers and researchers, for developing far more effective extension programs, and for ensuring that farmers get payments for their harvests that are market-related and permit long-term planning. Given secure landownership, proper market signals, and some training in sustainable farm management practices, even farmers on poor land can produce surpluses for growing populations in a sustainable fashion.

In many countries, large numbers of poor farmers require access to landownership. While a farmer with an adequate and stable income who owns his or her land will take care of it with a long-term view of its value, a poor farmer without an ownership interest cannot be expected to sacrifice maximum current yields for lower but sustainable output. Since women form the bulk of the agricultural labor force in many developing countries, land rights for women is a key issue.

Poor farmers also need greater access to credit so they can invest in improving productivity and sustainability. The infrastructure of roads and food storage facilities will require government investment. Without such an infrastructure, neglected in many countries, free markets cannot work. Farm-to-market roads and railroads, storage bins, processing facilities, research laboratories, and fertilizer plants are needed to support commercial farming. India, for instance, has greatly expanded milk consumption by increasing refrigeration facilities to reduce spoilage, rather than by trying to increase output.[21]

New irrigation schemes and better use of existing systems are crucial to intensifying land use. Roughly 17 percent of global cropland is under irrigation, and over half the increase in global food production since the mid-1960s is due to higher yields made possible by improved water control. Agriculture accounts for more than two thirds of the world's water consumption. Better technology and maintenance and better pricing of water to reflect scarcity would increase efficiency of use. Many irrigation schemes in developing countries have a delivery efficiency of only 30 percent because of water loss and waste.[22]

But solutions lie beyond technology and the internalizing of environmental costs: the most efficient irrigation schemes change the traditional dominating role of public agencies and rely much more on private management, research, and investment. Because of growing conflicts over water among farmers, households, and industries, however, the political process will have to set clear rules of the game for the market to play its part in allocating water more efficiently.

The modern farm in industrial countries employs few people. The owner/operator is highly skilled, running complex machinery, practicing sophisticated farm management techniques, and producing high yields of consistent quality. In many developing countries, however, this may not be the model to work toward. Labor-intensive agriculture, run by workers with better skills, is more feasible and may be socially desirable. But ways to achieve higher yields with minimal environmental impacts are still needed. Many techniques already exist, and many of them have been developed by farmers themselves. In the long run, developing countries will probably also want to shift to less labor-intensive methods of food production as their economies progress and as population growth slows.

Farmers in many developing countries cannot get enough chemical fertilizer at the right times. This is particularly true in much of Africa. Fertilizer use has been discouraged by depressed crop prices, scarce foreign exchange, high transportation costs, and government-sponsored monopolies that overcharge and distribute the product inefficiently. Farmers in poor countries also lack plant varieties adopted to make better use of plant food.

Agricultural chemicals can benefit developing countries when used properly according to local conditions. Africa suffers from cereal-eating birds and insects and from human diseases that deter agricultural development. Chemical sprays are conquering river blindness in Africa's Volta River valley by suppressing the flies that spread the disease. The introduction of predator insects to fight the cassava mealybug and green spider mite in African cassava is the world's biggest success with integrated pest management to date.[23]

The agrochemical industry has made great strides in assisting local training programs so farmers can improve food production without doing environmental harm, and its members are working with environmentalists, government officials, and international groups to develop a code of conduct for the marketing and application of agricultural chemicals. The U.N. Food and Agriculture Organization's International Code of Conduct on the Distribution and Use of Pesticides lays out the responsibilities of governments and businesses to help avoid chemical misuse in countries that lack good control procedures.[24]

Alley cropping (planting food crops between rows of trees) and other agroforestry techniques should be developed as a means of increasing food and fuel production on poor-quality lands. In any given year, about three fourths of Africa's arable land is left fallow, covered in young trees; alley cropping could thus quadruple the number of people supported by subsistence agriculture in the region. But currently it is limited largely to farm research stations. It will not be widely adopted by farmers until they begin to work with researchers to develop it for their own needs.

Ways Forward

The world agricultural situation holds important lessons for environmentalists and agriculturalists currently working in poor countries. Investments are needed urgently in the people, institutions, and infrastructure required to boost scientific and technical innovation in the farming sector, and farmers themselves must be equal partners in the innovation process.

Sustainable agriculture the world over requires greater efforts in several different areas to:

• open markets and define clear rules of the game to help business develop answers to existing problems;

• encourage increased food production on existing farmland and discourage the clearing of forests and wildlife habitats and the cultivation of fragile soils;

• encourage research to develop "best management practices" tailored to local farming conditions;

• support extension training for farmers;

• tie into and endorse GATT's efforts to eliminate trade subsidies and barriers distorting free agricultural trade; and

• support reasonable and largely incremental changes toward more sustainable development, rather than radical shifts.

Sustainability must be improved in developing countries. We do not have all the answers, and practices must vary according to local conditions, but we recommend:

• secure property rights and access to credit for poor and small farmers;

• more research and development of new technologies, plants and inputs best suited to fragile soils, and management practices tailored to poor farms;

• extension agent programs for farmers, and programs for farmers to train extension workers about the true nature of their problems, all of which must become more client-oriented;

• optimal use of fertilizers and crop protection chemicals;

• control of improper agricultural chemical use and an end to subsidies that lead to overuse of chemicals;

• improvements in most critical areas of infrastructure that will decrease food spoilage and increase availability; and

• restrictions on the farming of fragile soils and forests.

The world's growing population requires continued adoption of more sustainable agricultural practices. Such practices will enable farmers to make a living on their land without having to destroy important resources and ecosystems. And the earth will be able to supply abundant food indefinitely.

Forestry

About three quarters of the planet's forests have been brought into government ownership in the past few decades.[25] In most cases, government forestry has concentrated on providing raw materials to large industries, often in pursuit of export markets. But governments have rarely been effective at running forestry enterprises. Research has been poorly used, and governments lack adequate resources for optimal economic management of the forest, or even for controlling the use of forests by others.

Europe and North America suffered their greatest deforestation episodes in previous centuries. But it was not as disastrous as current tropical deforestation because temperate regions tend to have less fragile soils and to enjoy steady, year-around rainfall rather than short seasons of damaging, torrential rains, and because populations grew more slowly. Many Northern countries have found a new balance between agriculture and the area covered by forests; nations heavily dependent on the economic production of this sector, such as Sweden, now have more area in trees than 100 years ago.[26] The governments of some countries, such as the United States and Canada, however, are still criticized for poor economic and environment management of state forests.

Ineffective government management in many tropical countries has turned forests into economic vacuums into which the surrounding population flows—causing soil erosion, flooding, loss of species, and other problems. The task of keeping increasing numbers of people out of forests adds further to administrative burdens, decreasing resources for research and sound management. In fact, a number of developing countries encourage settlement following harvesting of forests.

Thus the national forestlands in these areas do not pull their economic weight. Their output is of little benefit to the surrounding rural population, who thus have little interest in cooperating with forestry authorities. In many communities of the developing world, forestry meets numerous basic needs: fuel, food, fodder, fiber, thatch, and other building materials. Thus many community groups have argued persuasively that local people should have a primary say in planning and managing forests for such uses, and that the role of government foresters is to help them in planning and management.

Governments have failed to generate adequate returns in almost all of Africa's state forest plantations.[27] The World Bank has concluded that better investment decisions tend to be taken by enterprises and individuals in competition with each other and with the outside world. There is ample evidence of thriving private-sector forest plantation development everywhere but in Africa.[28] Appropriate roles of the private sector include the establishment and operation of plantations for industrial wood production (subsidized on degraded land and marginal agricultural land), and production of industrial wood and nontimber products in tracts of natural forest zoned for production (especially secondary forest).

There is also scope for the private sector to join forces with local communities, through outgrower farm forest schemes (in which local people are paid for the wood they produce on their farms) and joint plantation ventures, in order to produce a wider range of products. The growing opportunities and responsibilities in forestry require private business to analyze the market conditions and the basic principles of sustainable use of this renewable resource.

Conflicting Demands on Forests

Wood is needed for all stages of national development. In developing countries, 80 percent of forest use is for energy, and sound forestry practices can help meet these energy demands. In industrial countries, wood is important for packaging and paper as well as for construction. In fact, a person in an industrial country is likely to use three and a half times more major forest products than someone in a developing nation.[29]

Forests also provide ecological services essential for economic progress. They regulate water supplies, moderate climate, protect soils, and provide a habitat for millions of species of potential economic value (having already provided the basis for many species now in cultivation). Over half the world's species are found in forests, but perhaps less than a tenth of them have been identified, and far fewer have been scientifically studied for possible useful application.[30] More than 120 compounds in the world's pharmacopoeia have been derived from plants, of which three quarters were found through ethnobotanical research—looking at the ways in which traditional societies use plants.[31] A mature forest is neither a net source nor a net sink for carbon dioxide, but growing forests do absorb this greenhouse gas.

Forests are homes for millions of people. They have been a primary source of development for countless cultures, and remain so today for many others. And they are a source of subsistence and welfare, especially for those without title to agricultural land and without formal employment. In more industrial countries, forests are also a source of recreation.

In using forests, however, both governments and private business have tended to focus more on the short-term material values of their products—usually the timber, fuel, or forest soil—to the exclusion of forest ecosystem and social services. This is because forest services rarely had a market value. As a consequence, developing-country forests have declined by nearly half over this century.[32] Deforestation can contribute to short-term economic growth and to the short-term alleviation of poverty, but it does so at the expense of long-term economic growth and other environmental and social goals.[33]

Forests now cover 30 percent of the earth's surface, less than half the area before settled agriculture began. And deforestation is accelerating—to between 17 million and 20 million hectares every year, 80 percent higher than a decade ago.[34] Most of this deforestation is in the tropics, where much of the highest-quality natural forest timber has already been removed from accessible regions. Of the 33 major tropical timber-exporting countries in 1985, only 10 are expected to be still exporting by 2000.[35] Comparatively few high-quality natural stands have not been exploited; these are in Zaire, the Guyanas, and some boreal forests. For every hectare of forest that is felled, less than a tenth of a hectare is replanted.[36]

In many parts of the world, especially the tropics, much of what passes for income-generating forest exploitation is simply the liquidation of forest capital. Where there are perceived to be "ample" natural forests to meet demands, and where governments make these available for cheap harvesting, investment in plantations will be inadequate, and logging will continue to degrade forests. This has created both social costs and costs to businesses involved in the forestry sector.

As a rule, the "renewability" of forests has been neither used to advantage by business nor kept intact. It is hardly surprising that members of the public in most forest countries are prejudiced against forest-using businesses. But when ecological value is respected and environmental costs are internalized, both social assets and the property of renewability—and hence profitability—can be maintained. Products must be harvested in ways that allow forests to renew themselves, which means retaining their ecosystem processes and social services. Under current systems this will be increasingly difficult, because demands for forest products and services are growing.

The demand for wood—the third most valuable major primary product entering global trade—will increase, although considerable uncertainties exist regarding how much. Projections range from increases of 33 percent to 75 percent between 1985 and 2030/2040.[37] The world's annual consumption of wood is already 3.4 billion cubic meters, the equivalent of a meter-thick, 85-meter-high wall stretching around the equator. Over half of this goes up in smoke, burned for energy.[38]

The global demand for industrial wood has increased gradually, at 1 percent per year, over the last two decades, with some of the highest rates of increase in developing countries, principally for construction. Yet 80 percent of the production and consumption of industrial wood remains concentrated in the temperate zone.[39]

Industrialization has brought a general trend toward reconstituted wood (particle board and fiber board) and wood fiber products, and away from solid wood. Reconstituted wood products can also take better advantage of recycling and waste reduction technologies, and can be made from a wider range of species. In the European Community, 50 percent of paper is now made from recycled fiber.[40] Recycling has its limits, however; 20 percent of wood fibers are destroyed with each cycle, and the energy costs of collecting and processing wastes can be prohibitive. Thus, large quantities of wood will continue to be required.[41]

But there is increasing flexibility in the type of wood inputs that can be used. The continuing shift toward reconstituted products, toward processing technologies that can use a wider range of raw materials, and toward increasing the efficiency with which raw materials are used and

> *"Deforestation of tropical forests is one of the most important of the global environmental protection issues. It is technically possible to reforest the globe, including badly degraded land. However, respect for ecosystems is critical. It is said that plantations destroy ecosystems because they are different from the natural vegetation, but they are an effective 'second-best' approach that supports the survival of the human species. Yet we must make every effort to sustain natural ecosystems as we reforest, and also to save resources and to recycle goods."*
>
> Jiro Kawake
> Chairman
> Oji Paper Co., Ltd.

reused will all continue to reduce the need for logs for a given end-use and to extend the range of available wood materials that can be used, according to British forester Mike Arnold.[42] The efficiency of processing can be greatly improved. Much of the natural forest resource is wasted: in West and Central Africa, half the wood logged is wasted, as is another half of that which is processed.[43]

Most wood products (except fuelwood in many places) now enter the marketplace, and this has ensured that increasing demands are met. In many countries, wood is grown commercially by farmers along with agricultural crops, and corporations are investing in plantations, which has encouraged commodity-based forestry research. This has improved economic yields, although it has been less successful in improving wood quality. But much remains to be done in terms of research, training, management efficiency, and technology cooperation.

Globally, expected increases in demand for industrial wood can continue to be met without appreciable real increases in prices, but there will be local exceptions, such as severe local fuelwood shortages in many poor countries. And in some industrial countries, consumers are beginning to take an interest in the source of their wood; they are prepared to pay more for wood from "sustainably managed forests" or to boycott wood from other sources. Such consumer reaction, once thought by the timber trade to be a fad, now appears to affect demand structurally.

Plantation forests can play a major role in meeting this demand, as many have exceptionally high growth rates, as much as 30 times those of natural forests. Although they account for less than 4 percent of total forest area, they provide about 20 percent of the world's industrial wood.[44] Planted forests consist largely of fast-growing species that need a lot of sunlight, notably pines, other conifers, and eucalyptus. Much of the investment in plantations and associated research into high-yielding varieties has been made by governments, but some of it has been made by corporations, particularly in high-yielding eucalyptus clones in countries such as the Congo and Brazil. About 100 million hectares of temperate plantations and 35 million hectares of tropical/subtropical plantations now exist.[45] The large plantations of Western Europe, Japan, New Zealand, Australia, Chile, Argentina, Brazil, and the southern and western United States will soon be brought into major production, but they will not meet global demands on their own.[46]

It is critical that increased demands on forest assets are met in ways that are sustainable. This will require:

• increased investment in plantations to meet projected demands, to restore ecosystem processes on degraded land, and, as far as possible, to make material use of degraded land as well;

• improved management of natural forests, especially of secondary (cutover) forests, for both production and ecosystem processes;

• protection of "primary" natural forests of highest conservation value, with no forest product exploitation; and

• forestry investments that honor social and environmental functions as well as economic ones.

Elaborating and implementing the principles of sustainable forestry and developing appropriate inventory and accounting techniques will require much more research and development, training, education, and management skills—and much more international cooperation.

Sustainable Forestry: Opportunities and Requirements for Business

Many characteristics of private business make it suitable for involvement in sustainable forestry: effective control of assets; financial resilience; the ability to use resources efficiently due to market competition; resources for afforestation; opportunities to combine forest investments with investments in forest-product-using industries and agriculture; good employment, training, and social development conditions; access to markets; access to technology and research capability; and institutional longevity.

One of the obstacles to corporate investment in forestry is the focus of modern capital markets on short-term profits. Planting trees will often have a negative impact on earnings. Secured long-term financing and proper valuation and accounting methods will motivate business to consider major plantation development for industrial wood. Historically, plantations have provided earnings of 4-5 percent per year, but many inefficient publicly owned projects have lost money, while private plantations have had returns of up to 20 percent.[47] Domestic markets have predominated until recently, but plantations (especially in the southern temperate zone) are increasingly being developed for export markets. Several large companies have long made forestry—both plantations and natural forest management—their principal business. This private forest industry is traditionally highly fragmented, with the largest company accounting for less than 5 percent of global sales, although American companies generally dominate. The activities of other companies are increasing.

Aracruz in Brazil is developing sustainable forestry plantations linked to a pulp facility. Some 7,500 employees work for this company in an area that 20 years ago was destroyed by deforestation—a development that also destroyed the local economy. The company works to combine technological and ecological know-how.[48]

The development of plantations may complement corporate investment in agriculture. Agriculture and forestry can provide mutual benefits, many of which are commonly unrealized in commercial concerns. Large U.S. agricultural concerns have planted extensive forest shelterbelts, which protect grain fields from wind erosion. Tobacco companies in southern Africa have planted woodlots to provide fuelwood for tobacco curing (finding it cheaper than bringing in coal). Commercial farmers in France grow poplars as part of the agricultural rotation. There are further, less commonly realized benefits: retaining natural forests can conserve the water source for agriculture, and serve as a reservoir of organisms that act against crop pests.

Establishing industrial plantations on just 5 percent of the 600 million hectares of land already deforested in the tropics would triple the tropical industrial wood harvest.[49] This implies a great deal of potential for joint business-government projects. The recently created private

Brazilian Foundation for Sustainable Development is preparing such "tandem projects" for rural areas, especially in the Amazon region.

Businesses have many possible roles in natural forests apart from the traditional enterprise of logging. Ben and Jerry's Homemade Ice Cream Inc. of the United States has developed flavors that feature Brazil nuts, cashew nuts, and assai and cupuacu fruits indigenous to the Amazon forests, where they are harvested by indigenous people.[50] The international cosmetics chain The Body Shop is importing Amazon herbs for use in cosmetics.[51] Both these efforts have taken special care to ensure that local people benefit as much as possible from the trade. Merck and Company, the world's largest pharmaceuticals company, has joined forces with Costa Rica's new National Biodiversity Institute to study forest plants for medicinal uses.[52] Costa Rica will gain payments for conserving its tropical forests, and Merck will gain exclusive rights to plants for screening.

Extractive reserves are now being set up in a number of tropical forest nations, and being pressed for in others. Under this concept, local people manage protected areas owned by the state in return for the right to harvest nuts, fruits, and other nonwood products commercially. If these are truly to benefit the communities concerned, significant local control of forest management, processing, and marketing will be necessary. There are still many unknowns concerning harvesting's effects on ecological balances, renewability of the resource, and the local society. Research is needed to ensure that extractive reserve production is sustainable.[53]

Land reclamation for other purposes also has commercial possibilities. The mining and industrial sectors are increasingly being required to ensure productive use of former quarries, spoils, and tips, and are using forestry to meet such requirements. Many large companies have their own guidelines, which anticipate future legal requirements, and some have made a point of leading the field. For example, BP's guidelines for sound environmental practice during oil exploration in tropical rain forests have influenced other corporations. The new environmental handbook of the Australian Mining Industry Council contains detailed recommendations on forestry for its members.[54]

Private forestry enterprises should strive to meet their commercial objectives while benefitting the local society, economy, and ecosystems. Occasionally the provision of social and ecological benefits from forestry can earn income, because demand for them enters the marketplace (recreation, water supplies, and land reclamation, for example), but more usually this is a cost, albeit a necessary one.

Sustainability requires an approach that treats social and cultural costs and benefits as seriously as economic and technical factors at all stages of plantation development. This approach is much more cost-effective if it is anticipatory. Getting the right mix will require good site surveys, concentrating as much on the ecosystem processes as on material resources. It will also require active consultation with local people. Tools such as environmental impact assessments and participatory rural assessments are helpful in this regard, and should complement the more usual detailed silvicultural investigations, market assessments, and cost-benefit analyses.

All types of investment in forestry—whether plantations, natural forest management, or forest processing industries—must be based on harvesting products at sustainable levels by careful control of harvesting levels and damage to residual stock and by monitoring and feedback into management systems. Sustainability also requires the maintenance of essential ecosystem processes by retaining continuous forest cover, returning nutrients to the soil through such things as in-forest debarking and conversion, minimizing soil compaction by the careful use of light machinery and animals, maintaining watercourse patterns, and careful control of chemical use. Also important is the maintenance of biological diversity at ecosystem, species, and gene levels through adopting multispecies/variety/clone systems wherever feasible, incorporating secondary succession where possible rather than treating the growth of other plants as a weed problem, using integrated pest management, and avoiding plantation development in natural forests.[55]

Tropical forest plantations may also include relatively large areas of land (20-25 percent) devoted to natural stands of native species both for social purposes and as a key element of an integrated pest management strategy. These areas may be left unmanaged or, after studies have revealed their important ecological roles, be upgraded with species or genotypes better adapted to handle environments and pests.

The needs of local people living in and around the forest must be satisfied by involving local people at all stages in forest boundary definition, planning, management, harvesting, and monitoring of the forest, and in forest product processing; by employing local people; by compensating them for rights and privileges forgone; by mutually agreeing access and usage rights; by providing recreation facilities; and by ensuring landscape and cultural compatibility.[56]

Investment in forestry research has proved profitable, but is often time-consuming. Breeding to improve tree species is a slow process, but genetic gains can be realized from even simple improvements in forestry. Choice of the right planting stock is one of the most critical factors in the profitability of the enterprise. More efforts must be devoted to such research.

Government Policies

Over the last decade, the roles of governments and local communities in forestry have been steadily redefined. Governments should protect forests of the highest conservation value, such as important watersheds and forests of exceptional biological diversity. They should manage natural forests that are subject to low-intensity exploitation (which will usually border those of highest conservation value). They should also establish a stable framework for more-intensive private forestry, support basic forestry research, and manage forest protection services (diseases, fire fighting, and so on).

Governments must recognize the comparative advantage of businesses for investing in sustainable forestry, and remove the obstacles that have made investment unattractive—unstable policies and markets, and poor access to land for adequate periods. Governments need to relinquish most of their roles in forestry production, which have proved to be inappropriate, and transform the production forestry sector into a market area promoting sustainable development.

For many businesses, the abundant market, social, and environmental uncertainties associated with long-term, land-extensive enterprises such as forestry can be compounded by the uncertainties of operating under the ambiguous regulatory systems of unstable political regimes. Thus credibility and consistency of the institutional political framework is of utmost importance. The general conditions must enable long-term investments and sustainable practices in forestry. This generally requires tax adjustments that highlight the capital appreciation and reduce the discount rates of investments.

The policy framework must especially:

• recognize the economic, social, and ecological values of forests;

• treat forests as renewable capital rather than stock resources;

• ensure business access to the right kind of land for forestry, which entails neither compulsory acquisition of private or communal property nor the clearing of natural forests for plantations;

• provide clear tenure, long leases and concessions, and effective land use zoning to avoid later conflicts with agriculture and other competing uses of land;

• allow developers to exercise management control of the forest over the long term;

• provide a climate of stability (low inflation, protection from high interest rates and undue taxes on land, stability in interest rates and taxes);

• provide clear performance-related regulations of conduct (the International Tropical Timber Organization has recently been developing standards for sustainable natural forest and plantation management); and

• provide a uniform political climate for conducting business, including free flow of finance, inputs, and trade of products.

Existing policy frameworks are generally more robust in industrial countries. Yet tropical countries have a strong comparative advantage for perennial crops such as trees, which can grow rapidly all year round in warm conditions and which require less cultivation of erodible soil than annual crops. Efforts to work with the governments of these countries in creating the right policy framework for forestry could be most rewarding.

10   Leadership for Sustainable Development in Developing Countries

"We recognise that sustainable development begins at home, that the costs of development must integrally include the costs of conservation, which, if not paid for now, will be extracted from the development process later or elsewhere."

Rajiv Gandhi
Prime Minister, India

The last three decades have shown an extensive rise in the growth of world production, trade, and per capita incomes. But average annual growth in per capita gross domestic product (GDP) for all developing countries declined from 3.9 percent over 1965-1973 to 2.3 percent over 1980-1989.[1]

World Bank forecasts for the 1990s were promising: annual per capita increases of 2.2 percent for developing countries, compared with 1.8-2.5 percent for industrial nations.[2] Good news was mixed with bad, however, according to early drafts of the Bank's *World Development Report 1992*. In South Asia, where about half the world's poor live, the share of the population below the poverty line was expected to fall from half to one quarter by the year 2000. Economic recovery in Latin America will also reduce the percentage of those living in poverty, but the number of poor in the region will stay about the same due to population growth. In sub-Saharan Africa, however, the number of poor people was expected to rise—by almost 50 percent during the 1990s due to rapid population growth and sluggish per capita income increase.[3]

Such figures prove that the "Third World" is no longer—if it ever was—a homogeneous place. Now that the "Second World," the communist bloc, has disappeared, the Third World reveals itself not as one region but as a spectrum of countries from the extremely poor to a few that are more "developed" than many countries in industrialized Europe. Each nation has its own preconditions and needs, its own path of development. Nevertheless, some basic concepts offer all countries guidance for the future.

First, as a rule, it is not lack of natural resources, finance, or human talents that are the main hindrances to economic development. As argued in the *World Development Report 1991* and the *Human Development Report 1991*, it has usually been domestic policies and patterns of resource allocation that have determined economic growth and development.[4] Second, economic growth is a necessary but not sufficient precondition for improved social equity and for more environmentally sustainable development. Third, unless the developing regions of the world, where 90 percent of future population growth is expected to occur, are put on a sustainable path, their problems will affect the more prosperous areas of the globe.[5]

Where governments have tried to carry out nation-building largely by themselves, with little reliance on the entrepreneurial skills of their people, the result has at best been wealth for a minority and relative poverty for the majority. Poverty is bad for human beings; it is also bad for business. That may sound callous, until it is remembered that it is business, in all its myriad forms, that recognizes human needs expressed in the marketplace and works to meet them in the most efficient ways. Thus business is concerned about mass poverty not because businesses can or should be effective charities, but because business can assist development by being more effective businesses. Among the policies of the poorer countries that entrench poverty are those that limit effective business participation in development, that limit the growth of markets, and that restrict the possibilities and thus the benefits of open trade.

Business leaders in the developing world are often accused of taking a short-term view. That they do so is hardly surprising, given the insecure investment climate. What is more surprising is the short-term views reflected by the development policies adopted by many Northern governments and the businesses that lobby and counsel them. These governments have often followed policies that have hampered development: trade barriers, a casual approach to the debt crisis, reduced and poorly directed aid flows, tied aid, and so on. This record should be kept in mind when reading criticism of the policies of Southern governments contained in this chapter.

The developing countries are seeking real economic growth rates of at least 4-5 percent a year. This growth is eroded by population growth rates of 2-2.5 percent in most of the developing world, but of almost 3 percent for the African continent as a whole. Real annual growth in the developing region of 5 percent would require countries to increase economic activity tenfold over the next 50 years, assuming that current economic practices prevail. They would then—in 50 years—approach the present economic level of the members of the Organisation for Economic Co-operation and Development (OECD). Many environmentalists and economists, however, predict a catastrophe of global proportions if such growth actually takes place.[6]

Most developing nations tend to seek economic growth by perceiving and exploiting many natural resources as cost-free input factors. This approach puts an immense burden on the environment and can generate huge environmental costs that must one day be paid. Many industrial countries also developed by exploiting natural resources as cost-free input factors. But they did it when their populations were smaller and resources more widely available on a per capita basis. They also brought resources in from colonies. Today, their growth occurs in less resource-intensive economies.

The newly industrializing economies (NIEs) of East Asia, especially South Korea and Taiwan, are witnessing pollution problems usually found in the early stages of development. They are experiencing development patterns typical of Britain in the late eighteenth century, the United States in the mid-nineteenth, and Japan in the early twentieth. But the scale of change is much greater and their environments much more vulnerable.[7]

It is frightening to imagine these same development patterns occurring in the more populous countries and regions. How can more than 1 billion Chinese, 820 million Indians, or some 600 million people of the African continent follow a development path based largely on natural resources such as soils, water, and forests—especially as all those populations are expected to double in the first half of the next century? The possible effects of climate change mean that developing nations cannot afford to base progress on intensive use of fossil fuels. This would be a disaster for them as well as for industrial nations. The developing world is thus following the development patterns of the industrial countries up a blind alley.

If they can no longer develop through intensive resource use, what are they to do? We must frankly admit that we have no complete model to offer. First, the history of "development models" has in hindsight been a history of development fads. Second, conditions vary greatly from country to country, region to region. Third, it is questionable whether the apparent success stories of the NIEs offer blueprints for others to copy. The NIEs benefitted from homogeneous, self-disciplined populations managed by governments with efficient bureaucracies that have shaped market economies successfully. Can these conditions be replicated elsewhere? The answer is at least not clear.

The search for new paths, which has only just begun, must be based on local initiatives and decisions supported by the utmost in international cooperation. It will need systematic learning from the best that industrial and developing countries have to offer.

The concept of sustainable development offers guidelines but no model. Its main truth is that economic progress, social progress, and the sound management of environmental resources must all proceed apace. To strive for the first goal while ignoring the second two destroys the basis of all progress.

Many leaders in developing nations are suspicious of this concept because they see it linked to a new "green conditionality" or even "green imperialism." They fear that OECD nations will dictate the ecological conditions under which aid will be given. Their suspicions of conditionality and green protectionism are justified. But they must begin, as many have, to examine the logic of sustainable development not as something imposed from the outside but as it applies to their own national situations and perspectives. Their main task, and that of businesses in the developing world, is to meet the needs of today without robbing the very next generation of needed resources—particularly in countries where the very next generation will be twice as numerous.

The only way forward is to "decouple" economic growth from environmental impact. This is being achieved elsewhere by technological change, by substituting resources, and by changing market signals so that the environmental costs of farming, forestry, the extraction of other resources, and pollution are all reflected in prices. Low, often subsidized, costs of raw materials have kept developing countries behind industrial ones in efficient resource use. Progress must be based on individual nations' own realities and possibilities, but it will also require international cooperation.

"The developing world has two options. The first is to sit back and react only when the problems arise, the way the developed world did. The second is to act as conscious global citizens and rise above our vested interests for the sake of future generations, so that history does not record that we deprived them of their livelihood."

Ratan N. Tata
Chairman
TATA Industries Ltd.

This raises the thorny issue of the conditions placed on things like aid, trade, and debt relief. Conditionality, a concept condemned by most developing-country governments, is daily bread for business leaders. They know that there are conditions under which they receive capital on the capital market. Yet in the business world, conditionality travels both ways; businesses expect their partners in the capital markets to meet certain conditions too. The substance and form of the conditions are decided by bargaining between the two sides.

Thus conditionality may be acceptable, as long as conditions are agreed to by both sides. A better expression might be "reciprocal commitments" or "mutual accountability," whereby both sides are accountable for their actions. Leaders in the developing world are having a hard time taking seriously the industrial world's commitment to sustainable development, as some of the richest nations find it impossible to agree to pollution limits and continue to increase trade barriers. Some of these will not discuss a change of life-style in their own countries, but instead plan cuts in their programs for cooperation with developing countries based on additional green conditionality.

Leaders of political, scientific, business, and citizens' organizations from both North and South America met in 1991 to explore the possibilities of new bargains. This New World Dialogue sees the possible need for a new global "social contract" for environment and development.[8] Such work points the way toward a new partnership between governments (which would define clear rules of the game and the business framework), business (which would contribute investment, know-how, and efficiency), multilateral organizations (which would provide financial and technical resources and advice), and nongovernmental groups (which would supply practical know-how). To reach sustainable development, each agent has its own field of action, but each one has also an opportunity for cooperation—and a duty to do so.

Obstacles and Opportunities

Preceding chapters have analyzed many of the universal needs and opportunities associated with sustainable development. Requirements such as internalizing environmental and social costs, appropriate market signals, more liberal trade, and more "sustaining" capital markets all apply to both industrial and developing nations. But at least six additional areas of concern are, from a business perspective, of particular importance in developing countries: population growth; poverty, migration, and the environment; external debt; ineffective rules of the game; small businesses; and education and training. The relative significance of each factor differs from one country to another, but their cumulative effect renders sustainable development difficult, if not impossible.

Population Growth

The world's population increases by 1 million people every four to five days. Of approximately 144 million children born each year, 126 million are in the developing world.[9]

A simplistic business view of this situation is that growing populations mean growing markets. But this is true only where needs are being converted into markets. In too many countries needs remain needs, and the poorest in their millions remain largely outside their nations' political and economic systems, and outside any but the most basic, usually informal, markets. It is these, the poorest, who are having the largest families—to help work the land, to provide social security in their old age, and to achieve social status, as children are the only "wealth" many couples are capable of producing.

The bottom line in considerations of population growth is that the only way to slow it is to first lower infant mortality rates. This truth—that to produce fewer people, more children must be kept alive—at first seems to contradict common sense. But families will have fewer children only if they know that the ones they do produce have a good chance of surviving into a relatively healthy and productive adulthood. No nation in this century has lowered population growth rates without first lowering infant mortality rates.[10]

The population issue is an intellectual, political, and ethical minefield. Every human life is inherently valuable, and there exists no single incidence of poverty, ill health, poor housing, or environmental damage that can be pinned specifically on "population growth"; the fault is more often found in policies. Environmental arguments for slowing human reproduction rates in the South pale before the truth that most environmental offenses with global impacts are committed by rich minorities with hazardous production patterns, energy use, and consumption styles. But there is no avoiding the fact that rapid population growth does overburden environmental resources. It also often overburdens economic progress, both of poor families and poor nations.

Almost 40 percent of the people in developing countries are under the age of 15, a fact that seriously constrains a society's savings ability and thus its ability to invest in physical capital.[11] Agricultural land is divided among growing families and marginal lands are brought into production. New jobs cannot be created rapidly enough. Health, education, housing, sanitation, and transport systems are overwhelmed. A slower rate of population growth is not in itself a guarantee of progress, because poor economic growth, poverty, and inequality have a multitude of roots. But poverty and population growth do reinforce one another.

The availability of contraceptive devices is important, as there is a tremendous and growing unmet demand in the developing world. But contraceptives cannot change social attitudes; motivation must come first. Infant mortality rates have fallen and contraceptive use has increased where governments have educated women, enlarging their options for self-fulfillment through participation in public and professional life, and have created a policy framework allowing people more access to basic needs such as food, land, housing, and health care. In some areas the demographic transition from a high mortality/high fertility society to one with low mortality and low fertility has been achieved in less than 25 years. Thus the primary responsibility for a demographic transition rests with the political authorities of developing nations.[12]

How can the business community help governments fulfill this responsibility? It can foster economic growth in a socially and ecologically appropriate way. Policy reforms that allocate more resources for primary health care, education, and social security systems are easier implemented in a flourishing economy. The business community can also have a desirable impact on the demographic transition by creating jobs and providing education and training for women and men equally.

Poverty, Migration, and the Environment

Population growth and poverty cause rural families to move, some into cities, some onto smaller farm plots, and some into areas that should be farmed not at all or only with great care and preparation, such as tropical forests or drylands. Both types of movement create massive environmental and social problems.

Surveys from 57 developing countries show that nearly half of farms are smaller than one hectare. In Indonesia, almost three quarters of farms are this size.[13] Many countries would do well to begin or accelerate programs to give farmers access to landownership. Given secure tenure, the right market signals, and the right information, many of the poorest farmers on the poorest lands can produce surpluses for growing populations and growing cities. Evidence from projects in Africa, Asia, and Latin America shows that yields from poor lands can be increased three or four times using only resources locally available.[14] Such reform not only helps keep farmers in the countryside, it should and can stimulate business growth and trade. Taiwan, for example, made it impossible to accumulate large amounts of wealth by owning land, so former landowners invested in industry instead.[15]

Land is more sustainably used where farmers have clear, firm property rights, which encourages them to care for their own natural resources, and where they have access to markets. Both preconditions are lacking in most developing countries. In Peru, one of the main reasons why as many as 100,000 farmers grow coca leaves for the cocaine trade is not a lack of profitable alternative crops, but the lack of property rights. Without them, they cannot get credit to develop alternatives. Also, the production of conventional crops is stifled by overregulation by the state. So, tragically and ironically, the illegal and therefore unregulated coca crop is the basis of Peru's only efficient, open, competitive agricultural market.[16]

Latin America possesses more than half the world's total rain forest reserves. Only 8.7 percent of the continent's surface is used for agricultural purposes, although more than 12 percent is thought to be ideally suited to intensive farming. Yet low-intensity cattle raising accounts for about 27 percent of the total farmed area.[17] Large "latifundi"—or big farms—throughout the continent keep productive land idle or underused. In the early 1980s, the large estates of northeast Brazil contained nearly 30 million hectares of unused or underused agricultural land; the situation is little changed today.[18] This is a terrible waste of land and of the productive capacity of the farmers kept off it. This sort of landholding pattern, and the associated lack of smaller farms and farming families, also prevents the growth of small businesses and of local market towns.

Migration from farming areas to cities and new agricultural frontiers is caused by push-and-pull effects of bad economic and social policies in agriculture and forestry, drawing young and dynamic people from their villages to what they see as the more attractive cities and virgin land.[19]

Land pressure is one of the forces pushing people into cities—cities that are already large and growing fast due to the population growth of their residents. Urbanization can be an engine of development, but only when it occurs slowly enough to be planned and guided and when those coming to towns are coming to jobs and bringing basic education with them. Urban populations of developing countries grow about 6 percent every year, approximately three times the rate of world population. In 1980, just 26 cities worldwide had more than 5 million inhabitants; by 2000, there will be more than 60 such cities, the "newcomers" being exclusively in developing countries.[20]

This concentration of population is hardly surprising, given that governments have established a policy framework that attracts people to the capital or the few other big cities. The best (and often most highly subsidized) health care, housing, sanitation, transport, and communications are there. Some of these factors help to attract business and industry, which in turn pull in those looking for jobs. Many studies have shown the negative sides of this urban squeeze: slums, high unemployment, crime, a deterioration in hygiene, and neglect of children and young people. Other effects include air and soil pollution, sewage problems, and a shortage of drinking water.[21]

Water is essential for life and health, but can also be a carrier of disease. According to the World Health Organization, most of the 500 million cases of the blinding disease trachoma could be eliminated if people in poorer countries could simply wash their hands and faces regularly with relatively clean water. In addition, water-related disease is responsible for half of child deaths and hospital admissions in the developing world.[22] Industrial countries were in similar situation until the nineteenth century. In Europe, cholera epidemics led to the construction of extensive water networks and treatment stations at the end of the last century.

Water quality is just as important as quantity. Rivers continue to receive and transport large amounts of wastes that may prevent their water from being safely used for drinking, farming, or industrial purposes. But water is scarce in large areas of the developing world, and women must spend hours collecting it from distant wells. In many big developing-world cities, few homes in the slums and shantytowns are connected to water pipes, and poor residents pay high proportions of meager incomes to water sellers. The better off, on the other hand, usually pay too little for the water piped into their homes.

Water issues thus epitomize the sorts of partnerships between business and government required for sustainable development. Governments must define political priorities and set the framework of managing water resources. They will usually find it cost-effective to provide public neighborhood standpipes for people who cannot afford to pay high prices for private water delivery; this will cut public health costs and improve productivity. But privatizing water delivery where public systems are unreliable and charge too little to those who can pay may also be an effective approach. Ideally, water prices should reflect pollution control and management costs. It is therefore desirable that water bills include, in addition to production and distribution costs, charges for pollution abatement and even taxes levied to finance infrastructure outlays. Once these costs have been internalized, the needs of the poor can be met by the sorts of credits and rebates discussed in chapter 2.

International corporations can be an excellent initiator of water technology cooperation, whether they are cooperating on massive water treatment projects or on the delivery of many simple, reliable village pumps.

It is obvious that poverty damages both the urban and rural environment. But the causes of this damage can be traced to the causes of poverty—all the many factors that hinder poor people from developing their talents and their potential. These include closed markets; lack of access to property, credit, and know-how; and weak education systems.

More-efficient urban environmental policies are both necessary and viable. Many public services (such as water and trash removal) could be privatized. Wealthy residents are rarely charged enough for such services, so it is no wonder that urban governments have trouble raising money. And governments should examine their overall policy structure to see which policies are fuelling urban growth. They should see what changes are needed to create opportunities in the countryside, with access to land being an obvious necessity in many areas. Provincial towns and cities need greater development based on an effective infrastructure. Opportunities in agriculture, forestry, and related industries must be improved as part of the attempt to reduce rural emigration. So must opportunities to start in business. Decentralization is a concept that has made successful entrepreneurs even more successful; this concept is pertinent to sustainable development.

Business has a vested interest in land use that is economically and environmentally sound, in privatized ecology-related services, and in effective cost structures. Business investment is essential if agriculture and forestry, together with their related industries, are to flourish.

Indebtedness

The combined external debt of the developing world at the end of 1990 was $1.36 trillion, a figure that had remained roughly the same for four years.[23] The interest payments alone on this debt account for about a third of total exports of the developing countries. About 60 percent of the total debt is owed to companies and banks (commercial debt), while the rest is owed to governments and international organizations (official debt). Most of it is owed by 17 "middle-income" countries, mainly in Latin America.

The absolute size of the debts of sub-Saharan African countries is small in comparison, but they form a relatively large proportion of the gross national products (GNPs) of these countries. Although progress has been made in restructuring the debts owed to official, bilateral agencies, there has been less success in dealing with debts to private creditors. About one third of sub-Saharan long- and short-term debts falls in this category, which means they are subject to interest rates averaging more than double those of official lenders.[24]

In fact, many African countries are unable to service their debt to private creditors and are accumulating even more. According to the World Bank, "this abnormal international financial situation weakens domestic confidence, makes macroeconomic management more difficult, increases the cost of trade finances and discourages possible alternative sources of external capital. It may also undermine the authority of debtor-creditor relations within the afflicted nations."[25]

Viewed from a business perspective, incurring debt is not negative if the credits are used with great efficiency and if priority is given to investment in development. In most developing countries, neither has been the case. Based on large amounts of liquidity in the international financial system following the oil price increase of the 1970s, the international banking community gave credits that were too generous to states that were too weak and too inefficient. The effect was disastrous because substantial portions of these credits ended up in public consumption, inefficient prestige projects, unproductive development schemes, corruption, and flight capital.

These debts, coupled with rising interest rates, have been a major impediment to economic growth and to reducing poverty and improving environmental conditions. David Knox, a former World Bank vice-president responsible for Latin America, has estimated that the debt service burden of that area needs to be reduced by 50-60 percent if a modest rate of per capita economic growth is to be restored.[26]

But it is surprising how quickly a country can begin to resolve its debt problems (without debt forgiveness) and improve its economy by changing the macroeconomic and political framework to cut public deficits, attract foreign and local investment, and encourage internal saving rates. This is basically what economically successful developing countries such as the Asian NIEs, Chile, and Mexico have done. Chile reformed its economic policies and institutional structure, and the country won back international and national confidence. With these came investment. Internal growth and the production of nontraditional exports—such as fruit and wood products—improved its capacity to pay its debts and at the same time gave it a stronger bargaining position in debt negotiations. Even flight capital that left some countries to escape bad governments and find more security in foreign investments was beginning to return. In 1991, the return of flight capital to all of Latin America except Brazil was estimated to have totalled about $15-20 billion.[27]

During the 1980s, there was a net flow of capital from the developing to the industrial nations, as repayments and the interest paid on outstanding debts exceeded development assistance funds and investments. By 1989 this had reached $50 billion per year.[28] In the early 1990s, this trend was reversed into a slight flow from rich to poor nations.[29] There are two main reasons for this reversal. First, there has been in an increase of aid and investment in Latin America. Second, some countries, such as Brazil, have not been paying the interest on outstanding debts.

The first trend is particularly encouraging. According to the London-based Overseas Development Institute, institutions such as the World Bank, the United Nations, the Inter-American Development Bank, and the U.N. Economic Commission for Latin America and the Caribbean are much more optimistic about the region's prospects than they have been since the early 1980s: "The reasons vary, but include the restoration of elected governments, implying popular support for new policies, changes in macro- and micro-economic policies, and the promises of more favorable US policies. The last, in particular, means that many regard the debt problem as, if not settled, at least on a well-defined negotiating path to resolution, even if its consequences remain a burden on past and future investment."[30]

In March 1989, U.S. Treasury Secretary Nicholas Brady announced a plan to reduce the bank debts of 38 debtor countries by 20 percent over three years. Banks were to accept new lower levels of repayment in return for a guarantee from the International Monetary Fund (IMF) and other international institutions that the reduced loan will be repaid. For their part, the debtor countries would have to agree to implement IMF structural adjustment programs and to make efforts to attract flight capital back and to promote foreign investment.

At its autumn 1991 meetings, the IMF proposed offering several countries relief on their official debt. More than half that of the Philippines ($15 billion) could be relieved in return for economic reforms. Similarly, Nigeria—Africa's largest debtor, with total debts of $32 billion—could gain relief on half its official debt if it accepted reforms, including a reduction in military expenditure. Proposals were also made to assist India, $70 billion in debt, and Brazil, which at $120 billion is the developing world's largest debtor.

Debt-for-nature or debt-for-sustainable-development swaps also offer some hope. This is essentially a mechanism whereby a debtor country exchanges a portion of its external public commercial bank debt for an amount of local currency equivalent to the secondary market value of the paper, plus a premium. The local currency then is used to finance projects related to sustainable development. The roots of these schemes lie in the fact that some developing-world debt can be bought at discount in the secondhand debt markets that have emerged in recent years.

By late 1990, $15 million had been invested by industrial countries and various conservation and development groups to purchase $95 million of face-value debt, which has been exchanged with the debtor countries for the equivalent of about $58 million in conservation and sustainable development funds and bonds. The funds support environmental programs in 12 countries.[31]

This process has had little impact on the debt problem overall, as it removed less than 0.01 percent of total debt. The swaps also appear to be self-limiting, both because of their inflationary effects in debtor countries and because if they are increased dramatically, recipients might feel that foreigners were gaining too large a say in running the countries.

Such schemes suggest the benefits of the "mutual accountability" approach, where both sides agree to obligations each must fulfill. The more positive relationship of accountability would give developing-country governments the opportunity to talk seriously with their creditors about the conditions necessary for success.

Ineffective Rules

Why is there less investment than needed in most developing countries? Why do many countries receive steady and growing capital inflow but the majority do not?

The attractiveness and reliability of the "rules of the game" of a given country determine its business climate and either encourage or discourage local and foreign investors. Once an atmosphere of trust and confidence has been established, new ways can be found to tackle the issues of growth, debt, and sustainable development.

"National prosperity is created, not inherited," according to U.S. professor Michael Porter. "It does not grow out of a country's natural endowments....A nation's competitiveness depends on the capacity of its industry (local and foreign direct investment) to innovate and upgrade. Companies (and with them the countries) gain advantage because of pressure and challenge. They benefit from having strong domestic rivals, aggressive home-based suppliers, and demanding local customers....Competitive advantage is created and sustained through a highly localized process. Differences in national values, culture, economic structure, institutions and histories all contribute to competitive success."[32]

The main elements of an attractive investment climate are known and proven: macroeconomic stability; free, open markets; clear property rights; and political stability. Unless these four conditions are largely satisfied, sustainable development is simply not possible.[33] This is why the structural adjustment programs of the World Bank and the IMF are to be welcomed (bearing in mind the points made earlier about the principle of mutual accountability); they increase the pressure on states to make the right changes.

Open and competitive markets have four major beneficial effects that are of particular importance to the developing world. First, they encourage competition, which leads to greater efficiency and innovation in the use of natural resources and energy. Second, they require unrestricted information and communications, which are essential for the transfer and increase of know-how. Third, they are a reliable indicator of acceptable prices to both producers and consumers; if prices also reflect the ecological cost of the product in question, both sellers and buyers will adapt accordingly. Fourth, they open opportunities to those who have talents and creativity; markets support equity of opportunity.

Domestic markets in most developing countries are subject to overregulation by the political system and public authorities. Prices are regulated both directly and indirectly; political influences at all levels of the economy manifest themselves in a veritable thicket of market regulations. Excessive and inefficient subsidies also distort markets and may encourage overuse of resources such as soil, water, forests, and agricultural chemicals. Access both to production factors (such as capital, land, and raw materials) and to markets is difficult and expensive. Worse, it is hard to assess the state of the markets because information is either lacking or filtered.

According to the Confederation of Indian Industry: "Recent research on competitiveness and relative industrial performance of different countries has demonstrated that the Indian methods of intervention are much less efficient than, say, Japan, Korea or Taiwan....While the inter-

> *"The new objectives demand open economies, democracy, and sustainable use of natural resources. The entrepreneurs' mentality has to change and adapt to this new vision. This is a great challenge for the present generation of entrepreneurs, especially in the developing countries, since it means that they must lead the way toward these ambitious new horizons."*
>
> Fernando Romero
> Chairman
> BHN Multibanco S.A.
> Inversiones Bolivianas S.A.

ventions in these countries reinforced competitiveness and efficiency, ours tended to impede efficient use of resources. There is therefore a rising crescendo of demand for greater use of the 'capitalist tools' such as the price mechanism and the fiscal and monetary instruments, rather than quantitative and direct controls. Some attribute Japan's dynamism to the fact that 'Japan is an economy driven by firms' and not the government."[34]   In 1991, India began to open internal and external markets.

A similar cry for open, competitive markets can also be heard from Africa and Latin America. According to Inter-American Development Bank president Enrique Iglesias: "We have to open up the economy....We have to restructure the economic role of the state. Countries have come to the conclusion that they are not limited to what the state can and should do."[35]

Thickets of rules and regulations cause massive underlying costs, or "transaction costs" as they are known. But transaction costs are often transaction profits for someone, and they make their way into the hands of a bloated, largely corrupt bureaucracy and those of the entrepreneurs protected by the system. New entrepreneurs are prevented from breaking into the market, and small businesses already active in the markets are frustrated.

The high cost of playing by distorting official rules forces most entrepreneurs to circumvent the system and sends many into the "informal sector."   In countries like Peru, Ecuador, and Brazil, this sector is estimated to generate more than one third of GNP, while the average for Latin America as a whole is about 20-25 percent. In many countries, the informal sector may provide more than 60 percent of all transport, construction, and retail trading.[36]

In many developing countries, the state's share of GNP is much higher than in many industrial nations, and economies are dominated by state-owned enterprises, which usually eat up far more in capital than the value they create. History continues to prove, in rich and poor nations, that the state is ill equipped to run businesses. But this is particularly so in developing countries. Not only are state enterprises economically inefficient, they are some of the worst culprits when it comes to environmental pollution.[37]

At the same time, however, there is an increasing tendency for governments to base environmental policies on command-and-control regulations, while industrial countries are beginning to rely on economic instruments. Such regulations, when administered by weak bureaucracies, decrease efficiency, increase corruption, and weaken government credibility. This is particularly true when, as in so many countries, state bureaucracies are inefficient, with poorly trained and badly paid staff. The government should focus on a minimum number of rules and regulations, but make sure that these are transparent and enforced.

Many developing nations lack a properly functioning legal system. Government policy may change from one day to the next, while presidents rule essentially by decree and are hardly accountable for their actions. Improvements in government should include a better perception by society of the government's commitment to improve general public welfare and be responsive to the needs of the people, its competence in assuring law and order and in delivering public services, and its ability to be equitable in its conduct, favoring no special interests or groups.

Small Enterprises

From a business perspective, one of the key ways of alleviating poverty in the developing world and spreading entrepreneurial talent is through encouraging the growth of small and medium-sized enterprises (SMEs). There are a number of reasons for this.

First, given that a high proportion of developing-world salaried jobs are in SMEs, encouraging these enterprises promotes opportunity in economic development. It creates jobs, income, and investment opportunities, particularly in rural and semiurban areas, where they are sorely needed. Rural industrialization generally involves small-scale activities with a low ratio of capital to labor. Hence, more jobs can be created for any given amount of investment. Second, SMEs are flexible, react quickly to needs and demands, and show talent for innovation. They form the seedbed for entrepreneurship, economic success, and social equity. With success, awareness for the environment increases.[38] Third, the large foreign debts of most developing countries have made international lenders, both official and commercial, hesitant about guaranteeing large loans to industries. There is a search for commercially viable activities in less capital-intensive sectors, and these can often be found in SME development.

In most developing countries, the nature and speed of agricultural expansion is one of the main factors determining how fast rural and semiurban industrialization grows. Incomes earned in agriculture are spent on local products. And as the agricultural sector expands, it uses inputs produced locally. This expansion in turn creates local industries to process its harvests. An agricultural sector that is not developing provides few linkages to support the birth and growth of small businesses. But these enterprises also face several other basic hurdles.

Small manufacturing enterprises typically employ 5-20 people; produce light consumer goods such as clothing, furniture, food (through baking and milling), and beverages for local markets; and offer services. They make a significant but often overlooked contribution to national economies. In countries with an average per capita income below $1,000, these enterprises account for 64 percent of all industrial employment.[39]

The most important hindrance for a dynamic SME sector is an overly complex, impenetrable, corrupt institutional structure. Capital is another major problem. Commercial local banks are normally not interested in this market because of relatively high administration costs for small credits and because of misconceptions about the dynamics and reliability of this market.[40] Informal financial sources such as money lenders serve small producers, but at interest rates three or four times above commercial bank rates. Therefore, small producers try to raise their capital through personal or family savings.

Small-scale entrepreneurs need access to credit schemes—to money and financial resources that would allow them to buy products in the marketplace and to pay back, through their earnings, loans they took out to invest in these small enterprises. The private guarantee scheme of FUNDES in Latin America suggests that private commercial banks would show more interest in this market if small entrepreneurs could offer better collateral in the form of ownership of houses or land secured by registered deeds.[41]

Small enterprises face great technological problems that cannot be solved by finance alone. Their productivity is severely constrained by lack of adequate technological and administrative know-how, tools, and equipment. This technology gap also hurts the eco-efficiency of SMEs. K.P. Nyati of India's National Productivity Council estimates that the small-scale industrial sector contributes about 50 percent of industrial output and 65 percent of industrial pollution. His conclusion was stark: "Small is beautiful? It can no longer remain so if the pollution from the small scale industrial sector continues to grow."[42] As the *Industry and Environment Review* of the U.N. Environment Programme (UNEP) commented in 1987, "small and medium size industries cannot tackle their pollution problems on their own; they need help, most importantly from large corporations that to a great extent have the necessary expertise and resources."[43]

The key task for improving the performance of SMEs is not research into new solutions but the diffusion of existing, useful information to those who need it. This requires the construction and reinforcement of information and cooperation networks from the global to the local level, meeting the needs of those involved in fostering small business. As part of its program to promote cleaner production techniques, UNEP's Industry and Environment Office has developed the computerized International Cleaner Production Clearing House. Its local "nodes," in the form of national or regional centers of cleaner production expertise, can disseminate information to the grassroots through existing or customized channels.

This decentralized network approach coincides with the conclusions of the BCSD workshop on harnessing small business for sustainable development: business associations are providing more environmental services for their members. International technology cooperation is especially useful in this field. Particularly interesting are associations' efforts to cooperate with SMEs on industrial estates or parks, industry-free zones, and export zones.

Education and Training

Education cuts through all issues of sustainable development in the developing world: Population growth can only be kept in check through better education, particularly of girls. Better land use requires better knowledge among farmers. Small entrepreneurs need training in administration and technical matters. More trained experts available locally will mean greater investment opportunities. Government bureaucracies need better-trained employees.

Investments in people are indispensable for sustainable development, technical cooperation, and much more. But such investments in the developing world have been modest—and in many countries, diminishing. Industrial countries spend about $2,000 per year for the average child's primary education. In developing countries with medium-range incomes, the figure is $200; in the poorest countries, it falls to just $20.[44] These ratios, far wider than the differences between the GNPs of industrial and poor countries, suggest that many developing countries put far too low a priority on the education of children.

Many developing countries also put too high an investment priority on tertiary education, favoring university education over both primary schooling and job training. This results in a manifest shortage of skilled workers on the one hand and a surplus of academics with limited job opportunities on the other hand. This frustrates people and frustrates business.

Investments by governments in education and training are the most critical factor in developing the capacity to innovate. Governments should give high priority among both economic and social policies to universal general education, literacy, and job-related and vocational training, as well as to postsecondary education. During the last 25 years many of the best and brightest from developing countries have received postgraduate education and research training in the universities of industrial countries, in a variety of scholarships and exchange programs. Many have emerged as academics and become involved in pure research rather than going into careers in industry, management, technology, and development. Many stayed to take up opportunities in the host countries. A participant in the BCSD Workshop on Technology Cooperation in India charged that many of those who did come home went into government laboratories "doing research of a theoretical or basic nature and not directed toward fulfilling any specific market need."

Governments should ensure that international scholarship and student exchange programs include enough students in business, industry, engineering, and management of technology to meet national needs for sustainable development and competitiveness. There should be exchanges of university graduates who already have some work experience and are employed by business and industry, to work directly in industrial research and development laboratories in other countries within university programs for advanced degrees. Universities should work with neighboring industries to operate and administer these programs of cooperative research and training. In some universities, this would be an extension of already highly successful cooperative programs with industry. The effect would be to bring businesses and universities together in cooperative work in both industrial and developing countries. It would provide exchange students with advanced degrees and direct industrial experience.

Training entrepreneurs in any country is a relatively new idea. There are many who believe that entrepreneurs are born, not made. Nevertheless, education and training in modern professional management methods and practices, especially in areas related to advanced technology, are essential for preparing future entrepreneurs for successful technology cooperation. Governments and business should therefore develop special programs of education for technological entrepreneurs.

Options for Business Leadership

Governments provide the framework in which business can get involved in the process of sustainable development. Recent world history offers striking examples of the fundamental importance of sound government policies for robust economic development. Business leaders and entrepreneurs around the world—in many developing countries in particular—are looking forward to improved conditions as governments begin to deregulate markets, privatize enterprises, and stabilize basic economic conditions.

But can business live up to its development responsibilities, given that it is usually seen as being mainly concerned with short-term profits, and often characterized as being interested only in the "fast buck"?

Business leaders, as well as politicians and indeed most people, have good reason to focus on immediate concerns: those who do not enjoy a relatively secure situation in the present are little motivated to care for the longer term future. So short-term success is important, and the hardships of developing countries give short-term profits a high priority there.

There will always be those for whom temporary success is enough. But those who are the leaders in societies—those in the position and with the desire to participate in shaping the future, defining strategies, policies, and structures—now face new challenges and new chances to combine their achievement of present success with efforts to safeguard the long-term survival of their societies.

This challenge is by no means new. People in many civilizations dependent on renewable resources such as forests, game, and fish learned long before economic theories were developed that they and their children can only be sustained by living off the interest of nature, not its capital stock. But this primeval challenge has reemerged in a formidably intricate form, reflecting the complexities of the modern global and technology-based civilization. Coping with such complexity will require the marshalling of much human skill.

Business leaders have special responsibilities and unusual opportunities in the global quest for sustainability.

They are called on to contribute their specific skills and experience to chart a new course and to define the new rules of the economic game, to muster all to join forces in their own self-interest. And they are expected to make this happen in the real world, where wealth is created and goods and services are supplied. Potential rewards are promising. Innovation and progress on the new course will bring substantial benefits in terms of new markets and increased market shares.

Making it happen swiftly and efficiently will confirm that the market economy, despite its deficiencies, is the best of available alternatives, and the only one on which a life in business can be based.

The true global business challenge, then, is to benefit from the system while contributing to and improving it. This is the essence of sustainable development.

CASE STUDIES

Successful Steps Toward Sustainable Development

As the first section of this book has demonstrated, sustainable development is a business issue that needs to be made a reality in each line function in every company. This section provides living examples of companies that have successfully met this challenge across a range of different issues. Already, hundreds of companies have begun to change course toward eco-efficiency. The 38 cases included here are a sampling of this broad-based movement within the business community. The companies featured come from all regions of the world, and include multinationals as well as smaller, locally based companies. They demonstrate that neither location, industry, nor scale are barriers to sustainable business practice, given the right external conditions and an internal commitment from both management and employees.

The 38 cases have been selected according to seven priority management themes: management leadership, industrial partnerships, stakeholder cooperation, finance, cleaner production, cleaner products, and sustainable resource use. As such, they cover the main issues that companies have to face when seeking to implement sustainable development practices in their everyday business operations. Nevertheless, there are issues that have not been covered. But sustainable development has to be seen as a process, and while the issues covered in this section are the priorities of today, it is certain that new issues will emerge as business leads the transition to sustainable development into the next century.

Similarly, the projects and companies illustrated are constantly improving their performance and fine-tuning their programs to take account of new needs and opportunities. What is presented here is thus a snapshot of current best practice; in each case, names and addresses are provided so that readers who would like to find out more can contact the company directly. Each case study describes how one particular company has gone about tackling a common sustainable development issue facing business. The cases are intentionally short, so that a flavor of the dynamics, results, and lessons can be obtained in the space of a few pages.

What emerges from these cases is ample evidence that companies can achieve commercial success while reducing their burden on the environment. Business's reputation for innovation can be successfully applied to the environmental arena through the development of new management techniques, products, and processes. The cases show how companies are moving toward concepts such as environmental auditing, stakeholder partnerships, and product life-cycle stewardship in real situations.

The examples unashamedly stress success rather than failure. Too often the media focusses on business errors and accidents, generating pessimism rather than enthusiasm for change. The following cases, then, are designed to provide positive examples to motivate others to take action.

Nonetheless, they are not models of perfection. In many instances, the cases show how the companies initially suffered environmental setbacks from a lack of foresight and commitment. They also show, however, that even those companies that have had a poor environmental record and reputation can improve their performance and their public image through effort and application.

Most of the cases were supplied by the companies concerned and then edited by BCSD staff. Unfortunately, the need for balance among business sectors, size of enterprise, and geographic regions meant that many good cases could not be included here. BCSD received unprecedented cooperation from companies, governments, and citizen groups in the editing process, a testimony to the collaboration required to make sustainable development a reality. Although BCSD has attempted to verify the facts in each case, it assumes no responsibility for errors of commission or omission.

We hope these examples will inspire others to find even better solutions in their own firms. There is no best solution. Each company must find its own course toward sustainable development.

Managing Change in Business

It is clear that good environmental management is not simply a matter of a government affairs department keeping abreast of legislation. It requires a response from every part of a company's operations and from all its managers and employees. This cannot be achieved in the traditional reactive manner, in which management interventions usually focus on fixing symptoms rather than on underlying causes. There are now simply too many forces and information signals from outside the company, and a reactive corporation is unlikely to remain competitive for long.

The imperatives of sustainable development imply that a business organization that intends to survive and prosper in the decades ahead must therefore become a "learning organization," built to adapt to rapid changes and to generate creative solutions more effectively than its competitors. This chapter considers five companies that have attempted to do this in various ways.

The oil price shocks of the 1970s forced many companies to reassess their basic business and adjust to changed circumstances. Case 11.1 describes how a U.S. utility company adopted an electricity conservation program that has become a profitable business and is producing significant environmental gains.

3M's Pollution Prevention Pays (3P) program is perhaps the best-known example of the benefits of pollution prevention. The company has worked to continuously upgrade and improve the system, as described in case 11.2. In the words of the current chief executive officer (CEO) L.D. DeSimone, "There's a good chance that what we do today will be judged by the rules of tomorrow. We must take the lead on environmental issues and set the pace of improvement ourselves."

In some industries, such as petrochemicals, environmental problems and rising levels of public distrust lead to the realization that the environment has become a central corporate issue. Case 11.3 looks at how Du Pont has responded to that challenge.

Case 11.4 examines innovative environmental auditing at Norsk Hydro. It demonstrates that companies see audits as an essential management information tool rather than a mechanism to police and punish managers who do not meet standards, and as part of a more cooperative effort to encourage improved performance.

The management, training, and motivation of employees is key to moving toward sustainable development. Case 11.5 describes how Shell trained employees in Nigeria, to the benefit of the company and local community alike.

Case 11.1
New England Electric: Making Energy Conservation Pay

Environmental and resource conservation issues are increasingly acting as one of the defining characteristics for many industrial sectors. For example, up until the first oil shock of 1973–1974, electricity utilities in the United States generated and sold an average of 7 percent more electricity each year.[1] Generally, the more electricity sold, the greater the profit. Then higher energy prices challenged the commercial wisdom of linking profitability to increased resource consumption. Growth rates fell, and consumer and environmental groups began to focus on saving rather than generating increasing amounts of electricity.

In the turbulent 1970s and 1980s, many utilities began to reassess their basic business, realizing that many of their customers wanted enhanced energy services rather than increased energy supply. Today, a number of U.S. utilities, including Southern California Edison, Pacific Gas and Electric, Duke Power, and Consolidated Edison, have introduced innovative programs to conserve electricity. One of the leaders in the field is New England Electric (NEE).

The companies of New England Electric collectively constitute a medium-sized utility, with annual revenues of approximately $2 billion, that serves the states of Rhode Island, New Hampshire, and Massachusetts. In 1979, CEO Guy Nichols decided his company should become a leader in encouraging electricity conservation and load management (now commonly called demand-side management). He had become convinced by the arguments of an increasing number of engineers and environmentalists that conserving electricity made good economic and environmental sense. Because the incremental cost of building new power plants was higher than the average cost of existing plants, demand-side management offered utilities the possibility of reducing capital investment programs for new generating capacity. In addition, less electricity generation meant less pollution.

New England Electric began pilot programs to help its commercial and small industrial customers conserve energy by investing in new lighting systems that were more energy-efficient; replacing old electric motors with new efficient ones and variable speed systems; and improving the efficiency of heating, ventilation, and air conditioning systems.

To start with, the programs were low-cost and experimental. Project proposals had to meet the "no losers" test (that is, the cost of programs should not increase the rates of nonparticipating customers above levels they would otherwise pay). The programs were important learning experiences for the companies, and met with considerable success: the public warmed to the prospect of reduced energy bills, environmentalists applauded the contribution to pollution prevention, and the company's public image was enhanced considerably. But the prevailing regulatory system linked profitability to electricity generation and sales, and therefore created a disincentive to further increasing conservation expenditures. Costs gradually rose from a modest research and development budget in 1979 to $40 million in 1988.

In 1987, while Samuel Huntington was CEO, regulators abandoned the "no losers" test. Huntington's discussions with Doug Foy of the Conservation Law Foundation (CLF), a leading environmental group, changed the way NEE and its regulators viewed demand-side management. New England Electric and CLF jointly developed programs and presented to regulators plans with steadily increasing investments. Although NEE and CLF had historically been adversaries on policy issues, both organizations realized the importance of energy conservation.

When John Rowe took over as president and CEO in 1989, the company had budgeted $40 million for conservation. Rowe recognized that demand-side management could only continue if the company could make a profit from its efforts. "The rat has to smell the cheese," he said. This meant changing the regulatory system. To do this, NEE again worked with the Conservation Law Foundation. Together they developed a proposal whereby NEE would agree to increase its investments in energy conservation and would be awarded a share of the generated savings. The regulatory agencies in Rhode Island, Massachusetts, and New Hampshire broke new ground in approving NEE's financial incentive for demand-side management.

The new program had four basic components:

• Customers would not have to pay for any conservation investment that exceeded the value of the cost of the electricity saved and the avoidance of building new capacity.

• NEE's earnings would be a share of the value created and would grow only as customers' benefits grew.

• The company would earn only when electricity savings were achieved.

• After a year, the estimates of savings would be open for public review and updated based on experience.

New England Electric companies were given financial incentives to encourage conservation among their customers. These incentives had the potential to add approximately 1 percent to NEE's return on equity. In 1990 NEE spent $71 million on energy conservation projects, cutting generation demand by 116.5 megawatts (MW) and saving 194,300 MW-hours of electricity. A total of $161 million was saved, yielding a net value of $90 million. Of this net value, NEE retained $8.4 million (9 percent) and its customers kept the remainder. Based on this initial success, NEE projected 1991 energy conservation spending to be approximately $100 million, 5 percent of its total revenue. This places the companies in the top tier of electricity utility conservation. In relative terms, NEE devotes a larger share of its revenues to conservation than any other U.S. utility.

New England Electric estimates it could spend $100 million a year on economically viable conservation projects within its service area between now and the end of the decade. NEE's continuing commitment to conservation is part of its overall environmental strategy to reduce its net air emissions by 45 percent between 1990 and 2000. These calculations do not include the possible emergence of new energy-efficient technologies during this time. Hence, the conservation opportunities and subsequent environmental improvement may prove to be even greater.

Although the new system has given NEE sufficient "cheese," to use Rowe's terminology, to invest heavily in energy conservation, there are now some losers. The unit price of a kilowatt-hour of electricity charged to all customers increased slightly to cover NEE's expenses, so customers who do not participate see extra costs.

Although the increase is smaller than it might otherwise have been with a new power plant, some larger industrial and commercial customers complain that they are being asked to pay higher electricity rates to subsidize conservation and savings by others. Using the inherent flexibility of the program, NEE is moving to address some of these concerns by increasing the financial contribution of participating customers when possible.

Lessons Learned

• Eco-efficient behavior often means changing the framework conditions to provide an incentive for change. For example, imaginative pricing schemes are needed to make energy conservation more profitable than building new power plants.

• For regulatory reform to occur, cooperation among utility executives, public interest groups, environmental groups, and regulatory authorities is essential.

• Energy conservation is possible and profitable.

Contact Person

Mark Hutchinson
New England Electric
25 Research Drive
Westborough MA 01582 USA

Case 11.2
3M: Building on the Success of Pollution Prevention

For business, sustainable development involves a process of continuous improvement in using fewer resources to satisfy consumer needs and generating fewer environmental impacts. Companies cannot afford to stand still. Management techniques and technologies are constantly being updated, and these innovations need to be progressively incorporated into existing programs and policies if a company is to stay ahead of regulatory requirements, community expectations, and its competitors. The U.S.-based company 3M, which produces a wide array of tapes and other consumer goods, has become justly famous for its pioneering Pollution Prevention Pays (3P) program; to meet the challenges of sustainable development, the company has now upgraded the program to produce the 3P Plus initiative.

The 3P program was introduced in 1975, making 3M the first company to develop an organized, companywide, multimedia application of the pollution prevention concept. The goal was to shift the focus from using traditional end-of-pipe pollution control equipment to preventing pollution at its source. 3P was based on the belief that the best way to prevent pollution is to not generate it in the first place. It has been successful because it encourages employees closest to 3M's products and processes to identify pollution prevention opportunities. Since 1975, more than 3,000 3P projects originated by employees prevented more than 1 billion pounds of polluting emissions and saved more than $500 million.

In 1986, Allen F. Jacobson, 3M's new chairman of the board and CEO, and Dr. Robert P. Bringer, staff vice president for environmental engineering and pollution control, saw that despite the success of the program, environmental issues were likely to have an increasing effect on 3M's competitive position in the marketplace. They determined that the growing volume and complexity of environmental regulations would make it difficult to react quickly to business opportunities. Further, it was evident that increasing costs and potential liabilities associated with waste and pollution would make it more difficult to maintain control of overall operating costs. Jacobson and Bringer believed that 3M must do more to prevent pollution and reduce the impact of environmental regulations.

To address these issues, Bringer brought to Jacobson and his operations committee a series of new environmental goals and activities to become known as the 3P Plus Program. The most important part is a long-term goal to cut all 3M emissions to the environment by 90 percent by the end of this decade. This marked a new stage in 3M's environmental policies, as it was the company's first formal goal for pollution prevention. To meet it, pollution prevention is being introduced into 3M business planning and into executive and employee performance reviews. The idea is to build on the success of 3P by using employee creativity, but to go beyond 3P by creating a more systematic approach to pollution prevention, with measurable goals.

The new program also includes a short-term goal to reduce air emissions immediately by 70 percent at plants worldwide by the end of 1993. This is being accomplished wherever possible by preventing pollution at major sources, even though this may not be required by law. However, the company recognizes that the public now expects lower pollution and that government regulations eventually may require it. By moving quickly, 3M satisfies a public desire and saves money by designing and installing its new equipment on its own time schedule.

Beyond the year 2000, 3M intends to improve its environmental performance and eventually achieve as close to zero emissions as possible. Other elements of the 3P Plus Program include a resource recovery initiative, new energy reduction goals, the phaseout of ozone-depleting chemicals, an upgrade of underground storage tanks, and the removal of polychlorinated biphenyls. 3M is on track to achieve its environmental goals. In 1990, air emissions were reduced by some 30 percent compared with 1989 levels; releases in 1991 were expected to be cut by a similar amount.

3M decided to maintain the popular voluntary nature of the original 3P program, but to bolster this with a more systematic approach so that the new targets could be achieved. The company is encouraging its managers to seek new products and processes that will both reduce pollution and improve product quality. However, it recognizes that the research needed to develop innovative technology can be expensive and time-consuming. Projects must compete with other company priorities for funding, and sometimes it is difficult to show that a 3P project will provide greater benefit than other kinds of projects. 3M scientists address this issue by attempting to link environmental improvements with product performance improvements or technological changes that provide business advantages. For example, the introduction of an innovative solvent-free technology eliminates the use of expensive drying ovens and results in lower manufacturing costs.

Bringer and 3M believe that top management commitment—especially of resources—"is the true proof of a dedicated management." Each year, 3M spends more than $100 million on research and development to look for innovative ways and new products to reduce its environmental impact. Without the commitment and motivation of its employees, however, 3M would have been unable to achieve as much as it has. "But with all of the demands on their time and energy, it is important to keep motivating and reminding employees of the importance of their environmental commitment and the fact that 3M strongly supports it," says Bringer.

3M has a system of awards for successful 3P projects and has used traditional methods for stimulating employees, such as meetings and company magazines. The company believes that direct financial rewards to employees can, however, be counterproductive by encouraging competition and secrecy rather than an open sharing of ideas. But financial rewards in the form of pay increases can result indirectly when successful environmental efforts are included in employee performance reviews.

3M believes that a key to its environmental success is the integration of environmental goals into 3M business plans. Summary environmental reports are submitted quarterly to the board of directors and monthly to the corporate operations committee. As Bringer says, "the long-term goal must be to make environmental management part of the company's culture." 3M executives are now required to incorporate environmental policies, objectives, and standards into their long-range planning, their research and development plans, and their capital and operating budgets. The company has also begun translating its corporate environmental targets into individual employee performance goals.

Lessons Learned

• Senior management commitment is essential, since it is senior management that sets priorities for the company, allocates resources, and motivates and encourages employees.

• Responsibility and authority must be assigned to an environmental officer at the highest practical level. A similar commitment must be made at each facility.

• Employees must be involved in the accomplishment of environmental goals. This can be achieved through top management support and formal recognition of successful employee actions.

Contact Person

David M. Benforado
Senior Environmental Specialist
3M
Building 21-2W-05
P.O. Box 33331
St. Paul MN 55133-3331 USA

Case 11.3
Du Pont: The CEO as Chief Environmental Officer

When he took over the chairmanship of Du Pont, the world's largest chemical company, in early 1989, Ed Woolard was concerned that the company's environmental performance did not match its standards in other areas. Although Du Pont's safety record had been outstanding, Woolard believed that corporate environmental performance could be upgraded. Also, the public still had a negative opinion of the chemical industry, viewing it as too secretive, potentially dangerous, and some- times arrogant. He wanted to instill a new sense of urgency concerning environmental issues throughout the company's management structure. "The basic problem was that management values were becoming out of phase with public expectations," says Woolard. "Although there were many examples of environmental excellence, they did not reflect a deeply held value of the company."

Woolard found that environmental management was largely compli- ance-driven and that environmental concerns had not been effectively integrated into other business areas. Many senior managers also downplayed the importance of environmental issues in their list of priorities. Woolard decided to take the lead by executive action. A marker was laid down within a month of taking office with his first public speech, at the American Chamber of Commerce in London, when he stressed that he was not only Du Pont's chief executive officer, he was also its "chief environmental officer." As such he put himself at the head of a new movement—"corporate environmentalism," which he defined as "an attitude and a performance commitment that place corporate environmental stewardship fully in line with public desires and expec- tations."

Woolard followed up this challenge to prevailing attitudes within Du Pont with a series of goals for improved environmental performance to be achieved over the coming decade. These addressed the major environmental concerns of the corporation, were quantifiable, and challenged managers to look beyond their own operations. They included:

• reducing hazardous wastes by 35 percent compared with 1990 levels, following a similar reduction program in the early 1980s;

• reducing toxic air emissions by 60 percent (1993 versus 1987 levels);

• reducing carcinogenic air emissions by 90 percent;

• eliminating or rendering nonhazardous all toxic discharges to the ground or surface waters;

• eliminating the use of heavy metal pigments in polymers;

• taking greater responsibility for plastic wastes disposal;

• including local communities in safety and environmental planning at all sites;

• providing for the management of 2,590 square kilometers of wildlife habitat, with a special emphasis on wetlands;

• phasing out the manufacture of chlorofluorocarbons by 2000 at the latest, and replacing them with safe alternatives; and

• procuring double-hulled oil tankers and double-walled storage tanks at all gasoline stations.

Woolard toured Du Pont, speaking directly to leading managers to help define the specific actions to be taken to implement these goals. A program of environmental seminars was started to heighten Du Pont's awareness of public expectations. Leaders of various environmental advocacy groups were invited to talk to all Du Pont employees. Meetings were also set up between environmentalists and senior management, at which areas of common interest were identified and projects sometimes initiated.

A number of organizational changes were also made to help institutionalize Woolard's vision and integrate environmental issues into mainstream business practice. The first task was to dissolve the elite executive committee in favor of a broader operating group, consisting of all senior management from Du Pont's operating businesses. Some personnel changes were necessary to build an executive team that shared Woolard's commitment.

Next, an environmental leadership council (ELC) was established, consisting of senior vice presidents from key businesses, functions, and international operations. The ELC determines policies and guidelines, and then reviews compliance within the corporation. By making this function the responsibility of line managers, rapid implementation has been facilitated. The ELC is served by a small corporate environmental affairs division headed by a corporate vice president, which has the dual role of tracking corporate performance and keeping the company abreast of the latest public concerns and scientific discoveries. This has enabled Du Pont to move from a reactive to a proactive approach to environmental management. For example, through its awareness and participation in discussions about global warming, Du Pont was able to announce plans to eliminate nitrous oxide emissions, an important source of greenhouse gas emissions, from nylon manufacturing.

To drive environmental excellence into day-to-day operating practices, environmental performance has been made one of the criteria for judging the compensation of senior managers. Senior and junior managers alike are evaluated for their compliance with company policy and legal requirements, their performance on corporate goals, and the impact of self-initiated programs. Du Pont also introduced an environmental awards scheme, Environmental Respect, designed to recognize individuals, groups, and businesses that have displayed superior environmental stewardship. A screening committee composed of Du Ponters and outside environmental advocates selects the annual winners from nominations. In addition to the award, winners receive cash grants that they can designate to the environmental concern of their choice. In 1991, its second year of existence, more than 600 projects were nominated.

Du Pont has seen substantive results in the area of attitude and motivation. Woolard's commitment to environmental excellence has served to legitimate the environmental concerns of many of Du Pont's employees. As Woolard himself has recognized, "I have never seen a stronger force for coalescing the organization about a common purpose than the environment. Our people were waiting to be empowered to do the right things." An important consequence of this enhanced sensitivity has been the development of a precautionary approach to environmental decisions, bringing together environmental concern and state-of-the-art science. Du Pont had shown its support for a precautionary approach with its decision in 1988 to stop production of ozone-depleting chlorofluorocarbons, long before governmental or intergovernmental requirements. In 1991, Du Pont announced plans to accelerate this phaseout to the end of 1996 in light of new evidence of faster than expected depletion of the ozone layer.

Du Pont has also achieved considerable progress toward its targets. For example, emissions of toxic air emissions fell by 30 percent through 1990. Archie Dunham, senior vice president of Du Pont polymers and chairman of the ELC, stresses, however, that the company is not resting on its laurels: "Much work will have to be done, and it is still too soon to feel comfortable that we will achieve all we've set out to do." Nevertheless, Du Pont has already achieved a certain degree of flexibility by setting itself tough targets, putting it ahead of regulatory requirements. Most of Du Pont's plants have established waste minimization teams, and innovative ways of managing wastes have emerged. In the company's acrylonitrile facilities, for example, waste reduction has gone hand in hand with lower operating costs. In a number of cases, what was once considered waste is now being sold as raw materials to other companies. Sometimes this has meant, according to Dunham, "that we now must produce these by-products intentionally."

New business opportunities are emerging. For example, Du Pont has found that its experience in minimizing wastes can help its customers manage their own wastes. In some cases, Du Pont has agreed to take back and recycle customer wastes free of charge. In addition, the company offers a range of environmental consulting and remediation services to non-Du Pont customers. Strategic alliances have been formed, built around environmental opportunities. A joint venture with Waste Management Inc. to build and operate a series of plastics recovery and sorting facilities, although no longer active, started Du Pont toward one of its corporate goals.

Du Pont's public reputation has also been improved, particularly through the growing commitment of local plant managers to improve the prospects for wildlife at their sites. Du Pont was one of the founders of the Wildlife Habitat Enhancement Council (see case 13.2), whose mission is to bring together companies and the local community to seek ways to use the huge areas under corporate control to maintain and sometimes increase wildlife.

Despite these successes, the company takes a pragmatic view of the future. "We do not know whether all the goals are truly achievable," says

Bruce Karrh, vice president for safety, health, and environmental affairs. "We are convinced, however, that we have started a trend within the company that could not be easily reversed."

Lessons Learned

• Committed leadership from a chief executive can unleash a cascade of environmental improvements throughout a corporation.
• Effort needs to be expended to develop understanding and ownership of the leader's vision throughout the corporation.
• Management systems need to be reformed and hard performance targets set to institutionalize a commitment to environmental excellence.

Contact Person

W. Ross Stevens III
Du Pont
Environmental Affairs Division
Nemours 11539
Wilmington DE 19898 USA

Case 11.4
Norsk Hydro: Environmental Auditing

Environmental auditing emerged in the United States in the late 1970s among environmentally intensive industrial sectors, such as the chemicals industry, as a way of assessing corporate compliance with a growing range of environmental regulations. It rapidly developed into an indispensable internal management tool for providing companies with vital information for their environmental risk management. Environmental auditing spread to Europe in the 1980s, largely as a result of U.S. multinationals wanting to assess the performance of their subsidiaries. It first took root in the Netherlands, then moved to Scandinavia and the United Kingdom.

Environmental audits to date, unlike financial audits, have been wholly voluntary, adopted by leadership companies to help improve their performance. The International Chamber of Commerce has defined an environmental audit as "a management tool comprising a systematic, documented, periodic and objective evaluation of how well environmental organisation, management and equipment are performing."[2]

The vast differences in corporate culture and structure mean that auditing programs need to be individually tailored to company needs. There is no one right way to audit, and no limit to what can or should be included in an audit. As the practice has developed, it has extended from mere compliance auditing to assessment of management systems, company responses to specific environmental issues, potential acquisitions, and, more recently, suppliers. Experience has also shown that an environmental audit is not a panacea for ending environmental abuse. Rather, it must be used as part of an integrated environmental management system.

Auditing is an evolving concept. Although it started as an internal management tool, there is increasing interest from governments, investors, unions, and the community to establish external access to audit results, and also to institute common auditing standards to ensure that audits are of high quality. A number of companies have pioneered corporate reporting, using their environmental auditing programs as a platform for informed dialogue with stakeholders.

As the environmental management journal *ENDS* noted in 1990, "the second half of the 1980s was not the happiest time for Norsk Hydro's reputation in Norway."[3] As a chemicals-to-energy conglomerate, and as the largest industrial employer in Norway, Norsk Hydro has a very high public profile. It also had a series of environmental incidents ranging from the discovery of extensive pollution of groundwater beneath its vinyl chloride plant in Rafnes in 1985 to a serious fire at the same plant in 1988. In addition, environmental activists discovered that soil around the Heroya chlor-alkali works was impregnated with mercury; furthermore, extensive mercury emissions were found at the Porsgrunn chlorine works, now closed.

To counter these problems and restore public confidence, Norsk Hydro decided to extend its existing health and safety auditing program to the environmental field. Rolf Marstrander, then vice president for health, safety, and environment, explains the company's philosophy: "We do it to learn, not just to find out what is wrong." The company has drawn up a plan that states that all companies and major plants must be audited every other year. Norsk Hydro has already learned a great deal from its auditing program, and is constantly perfecting its auditing techniques as it gains more experience.

Today at Norsk Hydro, internal environmental auditing is a standard management tool used both as a scorecard for current activities and as a method for planning, implementing, and following up specific projects. "Environmental management audits give rise to questions and issues that a company ignores at its own peril," says Per Arne Syrrist, acting head of Norsk Hydro's health, safety, and environment staff. The audit is now a full part of the management system, which allows senior executives to keep up with how the organization's activities are structured and operated, enabling them to raise overall environmental performance continuously.

Management decided, however, that internal auditing was not sufficient. The company's environmental performance had to be effectively communicated to the public. As a result, Norsk Hydro has embarked on a program of environmental reports, starting with its operations in Norway. According to Torvild Aakvaag, chief executive, "we believe that the public has a right to information. If we have a problem, it is in our best interests that it is brought out into full public view." The Norwegian report was circulated to Hydro employees and their families, to local and national authorities, to homeowners in the vicinity of Hydro sites, and to schools and colleges.

This was followed by a similar report detailing its operations worldwide, released in 1990. The report showed that emissions from all Norsk Hydro's plants were reduced by around 90 percent during the 1970s and 1980s. This success led the company to state that its Sluiskill fertilizer plant "is probably now the fertilizer production plant with the world's lowest emissions relative to production volume." By adopting a policy of openness and candor, Hydro believes that it has been able to regain the confidence of the public, reflected in its award as "the best company to work for" in Norway in 1990.

Although Norsk Hydro has now established an international auditing and reporting program, its decentralized character has provided opportunities for its subsidiaries to experiment with new environmental management ideas. Norsk Hydro UK Ltd. decided in 1990 to publish its own report. Although Norsk UK is a top producer of polyvinyl chloride compounds and the second largest manufacturer of nitrogen fertilizers in that country, the company was almost invisible to the public eye.

Environmental reporting was seen as a way both of communicating with the public about an important issue and of raising the public profile of the company. What made the U.K. initiative so innovative was that management decided it would publish the contents of the audits, whatever the findings, and that they would commission an independent third party to verify the audit results.

The audit report drew on the internal environmental reports prepared each month by U.K. plant managers and on the company's regular safety audits. Charles Duff, corporate development manager for Norsk Hydro UK, explains that the company wanted the findings checked by an objective authority, "to forestall any responses that we were publishing only part of the story—the part that was flattering to us." Rather than using a regular financial consultancy to conduct the assessment, Norsk UK chose consultants from Lloyd's Register. Managing Director John Speirs doubted "that accountancy firms are the right organizations to do it. They have a financial bent and I felt it was better to go to a firm whose consultants had more relevant backgrounds."

The 28-page U.K. report was published in October 1990 and was advertised in the business press, inviting readers to send for copies. So far, there have been more than 700 requests for the report from 24 countries. In addition, some 10,000 copies have been sent out to a variety of organizations, including the European Commission and Ministries of the Environment, and to Members of Parliament and chairs of companies. Taking a "warts and all" approach added significant credibility to the report, believes Duff: "We are continually asked to speak about our experiences, both at conferences and also to individual companies—some of them household names—who are looking for our advice about how to prepare their own audits." He adds that "the culmination for us in the UK has been the appointment of the Managing Director of Norsk Hydro UK Ltd to the Advisory Committee on Business and the Environment set up by the Secretaries of State for the Environment and for Trade and Industry."

But publishing environmental audits and the use of third-party auditors for verification is only one element of Norsk UK's environmental strategy. "External assessors have a part to play," says Duff, "but it is ourselves who know best what is actually going on at a site, who know best where to find areas for improvement, and have the authority and the commitment to bring about improvements."

In Norway, Norsk Hydro has now established a leadership position on environmental reporting, thanks to its 1989 and 1990 reports. It stands by its decision not to have its reports externally assessed. Syrrist argues that the company has 25 years of experience in offshore oil and gas operations in the North Sea, and in this industry Norwegian legislation directs operators to define, and monitor against, their own standards, within a legally required framework of procedures. Hydro follows the same approach at its onshore operations, and believes that its legal obligations to report its performance to the government are sufficient for the moment. The company has decided to continue publishing environmental reports, although not necessarily on a yearly basis. These will be supplemented with further local and regional reports.

Lessons Learned

• Internal environmental audits can help improve corporate environmental performance.

• Such audits provide a basis for reporting this performance to company stakeholders.

Contact Persons

Per Arne Syrrist   Charles C. Duff
Norsk Hydro   Corporate Development
Bygdoy Allé   Manager
0240 Oslo 2, Norway   Norsk Hydro UK
                                          Bridge House
                                          69 London Road
                                          Twickenham TW1 1EE, UK

Case 11.5
Shell: Human Resource Development

Foreign direct investment from a multinational corporation is often the most effective way of exchanging the skills and technologies needed to further sustainable development in developing countries. In particular,

foreign investors can contribute directly to the building of local management expertise and employee know-how through training programs.
This has benefits not only for the company concerned, but also for the
wider community. Managers and workers who later leave the company
to work for others or set up their own businesses will take the skills with
them, thus diffusing "best practice." This has been the experience of Shell
Petroleum Development Company of Nigeria.

In 1956, Shell discovered the first commercial oil field in the delta of the
Niger River. At that time, the country had very few personnel with skills
in the specialized aspects of oil and gas exploration and development.
Indeed, technical skills were in short supply generally among the Nigerian population. Initially, all management and virtually all skilled production employees were expatriates. Job opportunities for Nigerians
were limited to low-skilled and peripheral support positions.

Production levels rose from a modest 6,000 barrels of crude oil per day
in 1958 to more than 1 million barrels per day by 1973. Shell was well
aware of the high costs of maintaining expatriate workers. Their salaries
are almost always higher than wage rates for comparable local employees. In addition, there were high costs for travel, home leave, special
housing allowances, and family education and health care. Shell's management recognized that full participation of qualified Nigerians in all
phases of the oil business would be economically beneficial to the
company as well as to individuals employed and to the local economy
in general. From a political point of view, employment of Nigerians was
also an important economic development and social goal of the Nigerian
government.

To improve the supply of technically trained engineers from local
universities and technical colleges, Shell introduced a program of scholarships for secondary schools and for local and overseas universities and
technical colleges. In addition, it established its own directly funded
Technical Trade Schools. Shell managers and the company's Nigerian
government partners also decided on specific targets for Nigerian employment. Both parties agreed that Nigerian employees would meet the
same high technical standards as applied by the company in all its
operations throughout the world. Quality and safety standards would
not be compromised, and good environmental management would be
enhanced.

The results of this long-term commitment to education, training, and
technology cooperation can be seen in table 11.1.

Table 11.1
Employees in Shell Nigeria

Employment Category	1970		1990	
	Nigerian	Expatriate	Nigerian	Expatriate
Exploration/ Development	122 (46%)	142	543 (87%)	83
Oil & Gas Operations	136 (45%)	165	623 (94%)	40
Support Services	121 (59%)	83	911 (96%)	36
Management	0	12	13 (62%)	8
Total	379 (48%)	402	1,530 (90%)	167

It is not surprising that by 1970, Nigerian employees already exceeded the number of expatriates in the support services category. But it is significant that just 12 years after production began in 1958, Nigerians occupied 45 percent of operations jobs and 46 percent of exploration positions. As Godwin Omene, now Shell Nigeria's deputy managing director and a former Shell scholar at Imperial College, London, says, "the commitment to education and training generated results in a relatively short time frame." However, it took longer for Nigerians to develop enough broad-based specific oil industry experience to move into management positions. But by 1990, Nigerians held 62 percent of the management positions, including the deputy managing director, the general managers of administration and of exploration/non-traditional business, and the divisional general managers and operations managers in Port Harcourt and Warri.

During the 1980s, the company increased its efforts in training, both local and overseas, significantly. Shell also assigns its Nigerian employees to other Shell companies worldwide for up to five years. Currently, 70 Nigerian employees are assigned to Shell operations outside that country.

As a result of Shell's training and education efforts, 90 percent of all Shell jobs in Nigeria are now held by locals. But this tells only part of the story. Former Shell employees have taken important positions with government agencies or other private firms or have established their own businesses built around the experience gained at Shell. Indeed, businesses have been established by former Shell employees in engineering, construction, oil exploration, computer consultancy, drilling, accountancy, and legal practice. B.A. Osuno, director of the Ministry of Petroleum Resources, S. Ihetu, managing director of the Nigeria Liquified Natural Gas Company, C. Ezeh, managing director of the John Holt Group, and Jimi Bademosi, managing director of the Nigeria Gas Company, are all former Shell employees.

The training program thus has a ripple effect through the Nigerian economy by providing a larger pool of technically trained personnel. Without this expanded pool of skilled technical staff, technology cooperation in many fields other than oil exploration and production would be more difficult to achieve.

This complementary process has its limits, however. Shell in Nigeria had to cope with the problem of higher-than-normal turnover rate during the late 1970s because of the rapid expansion of the Nigerian economy and the resultant opportunities for well-trained personnel in the wider environment. Functions (such as computing, finance, and field engineering) that were not limited to the oil industry were most affected.

Lessons Learned

• Investment in technical training pays dividends for the company, for the individuals involved, and for developing countries in general.

• The localization or regionalization of corporate staff can occur without compromising corporate quality or safety objectives.

• Investments by multinational corporations can make significant contributions to local development objectives.

Contact Persons

P.G. Tauecchio   N.A. Achebe
Human Resources Adviser   General Manager,
Exploration Staff   Administration
Shell Internationale Petroleum   Shell Petroleum
  Maatschappij B.V.   Development Company
Postbus 163   PMB 2418, 21/22 Marina
2501 AN The Hague, Netherlands   Lagos, Nigeria

12

Managing Business Partnerships

Progress toward sustainable development will require innovative business partnerships. Technology cooperation between firms, with benefits for all parties, is an effective way to enhance both productivity and environmental quality. Working with suppliers to ensure quality control and improved environmental performance can help purchasing firms as well as the suppliers. Industry trade associations can also act as catalysts to help an entire industrial sector.

This chapter looks at three types of corporate partnership: technology cooperation between companies in industrial and developing countries, partnerships with suppliers, and partnerships with other companies in the same industry sector.

Case 12.1 describes a long-term collaboration between the Japanese and Brazilian governments and between companies in those two countries in the construction of an integrated steelworks, showing how different types of technical and financial cooperation were appropriate at different stages of the project. A great deal of technology cooperation will be needed to help rebuild Eastern Europe, and case 12.2, on ABB Zamech, looks at how a joint venture in Poland helped improve product quality while improving environmental performance. Case 12.3 considers a corporate decision to phase out the use of a potentially hazardous material, and the cooperation between Swiss Eternit Group headquarters in Europe and group companies in Central America to find alternative substitutes for traditional products.

S.C. Johnson Wax has extended an existing supplier relations program, put in place as part of an ongoing quality program, to serve as the basis for addressing sustainable development issues. Case 12.4 features an S.C. Johnson special conference to educate suppliers and communicate environmental goals, as a first step in forging action-oriented partnerships to improve environmental performance.

Companies within an industry sector can work together to improve their environmental performance, as described in case 12.5 regarding the chemicals industry. Case 12.6 looks at how an industry trade association in Kenya is helping the highly fragmented leather industry become more eco-efficient.

Case 12.1
Nippon Steel/Usiminas: Long-Term Partnership for Sustainable Development

Multinational corporations have long been the primary agents of technology transfer to developing countries. In many cases, the technology concerned has been used both to foster development and to improve environmental standards beyond those in the host country. Technology cooperation, however, involves far more than the simple exchange of production hardware. It encompasses a long-term partnership based on the total quality concept of constant improvement.

Experience has demonstrated that technology cooperation works best when both parties have mutual and long-term interests. Training is a critical element in the success of any such cooperation. It is also essential that the favorable economic conditions are established through the creation of a positive investment framework in the host country that encourages companies to invest in and develop local resources. For more than 30 years, Nippon Steel has been practicing this philosophy through its joint venture with the Usiminas steel company.

The starting point for the partnership was the Brazilian government's desire to industrialize, using its abundant natural resource base. It asked the Japanese government for assistance in developing a basic steel industry in the state of Minas Gerais. The Japanese government agreed to provide low-cost loans for the initial investment, and asked the Keidanren trade association and the Japan Iron and Steel Federation to provide technical expertise. Of the original capital investment of $191 million, 14 Japanese companies contributed $41 million through Nippon Usiminas. The Brazilian government also issued low-cost loans for the project and gave import tax exemptions for some of the original equipment.

The joint venture, Usiminas, was established in 1958 as a state-controlled company; in 1991 the company was privatized. Two steel companies, Nippon and NKK, provided technical assistance in the design and construction of a "greenfield" steel plant with a capacity of 500,000 tons per year. This plant came on-line in 1965. Usiminas subsequently expanded its capacity to 2.4 million tons in 1976, and then to 3.5 million tons by 1982.

To facilitate independent and efficient operation of the plant by the Brazilians, a number of measures were taken. Japanese personnel were assigned as managers at Usiminas in order to train their Brazilian counterparts. The largest number of Japanese serving in these positions at any one time was 130. A number of key people from Usiminas were also sent to Japan to receive extensive training at Japanese steelworks; 10 key staff worked and were trained in Japan for one year. The handover from Japanese to local management took about two years.

Subsequently, Nippon Steel took on the roles of engineering consultant and advisor on operational know-how. Technical assistance was provided for the expansion of capacity and the transfer of technology. During the height of the expansion period, about 10 Japanese were sent annually to Usiminas. Today, no more than two or three are needed due to the increased in-house capabilities of their Brazilian partners.

Since steelworks are large-scale production facilities that consume vast quantities of raw materials and energy, creating various types of pollution, environmental control was a part of the project from the start. During the 1960s and 1970s, tough environmental protection legislation in Japan forced considerable technical developments in the steel industry. In turn, these improvements were built in at the Usiminas plant. A wide range of environmental control equipment has been installed, not only to comply with Brazilian laws, but also to establish a good living and working environment for local people. Usiminas's total investment in environmental control management since 1962 has amounted to $190 million—approximately 7 percent of the total investment in the works from 1962 to 1991. In addition, Nippon Steel sent a number of environmental protection experts to Usiminas between 1979 and 1981 to conduct on-the-job environmental training, while a number of engineers from Usiminas were sent to Nippon Steel plants. These exchanges have continued until very recently, and were mostly paid for by the Brazilian partners.

The main improvement to the local environment came from the introduction in 1981 of an oxygen gas system used to clean and recover basic oxygen furnace emissions from steel mills. Brown smoke from the plant has virtually disappeared, with dust fall in the areas around the steelworks dropping by 90 percent from pre-1980 average levels. The creation of "greenbelts" around the works has served as a natural filter that absorbs dust and noise. At present, 2.6 million trees have been planted and are growing in a 1,992-hectare afforestation zone around the Usiminas facility. The company is now spending an additional $58 million to further improve the environment in the area immediately around the steelworks.

The high energy consumption inherent to steel production means that Usiminas is engaged in a constant effort to cut energy use. The introduction of the continuous casting process, for example, let the company eliminate a number of intermediate processes in forming steel products from molten steel. In comparison with conventional processes, continuous casting can save 150,000–200,000 kilocalories (kcal) of energy per ton of crude steel. Usiminas now produces 86 percent of its total output by continuous casting. In addition, Usiminas has reduced fuel consumption in its rolling mills by about 20 percent from the 1979 level by intensifying the combustion control of the heating furnaces.

The level of energy consumption per ton of crude steel in the Brazilian steel industry is almost the same as that in Europe, and much better than that of the North American steel industry. As a result of energy conservation measures at Usiminas, energy use per ton of crude steel has remained stable at about 6 million kcal for the past 10 years; however, the consumption of electricity has had to be increased in order to operate dust collectors and other environmental control equipment.

The Brazilian government has succeeded, with the cooperation of Japanese partners, in establishing a world-class steel company in terms of management efficiency, productivity, and environmental protection. Steel is now a major export item for Brazil and, as such, a major earner of foreign exchange. Data from the Instituto Brasileiro de Siderugica indicate that the country earned $3.6 billion in 1989 and $2.8 billion in 1990 from steel exports. For Nippon Steel, the benefit has been the establishment of an excellent international reputation in the field of steel engineering.

Lessons Learned

• Technology cooperation between corporations in industrial and developing countries is possible and profitable for all concerned, if based on mutual trust and favorable business conditions.

• A continuing partnership ensures that new innovations and environmental improvements will be introduced quickly in developing countries.

Contact Persons

Makoto Yoshida	Ralfe W.M. Nogüeira
General Manager	Coordenação de Assistencia
Environmental Management	Técnica OAT
Division	Usinas Siderúrgicas de Minas
Nippon Steel Corporation	Gerais SA
6-3 Otomachi, 2-chome	Usiminas
Chiyoda-ku	Belo Horizonte, Brazil
Tokyo 100-17, Japan	

Case 12.2
ABB Zamech: Technology Cooperation Through Joint Ventures

The collapse of communism in Eastern Europe has revealed an industrial sector lacking modern technologies and management techniques, where an absence of effective competitive or democratic pressures has resulted in economic inefficiency and large-scale environmental abuse. The region now has a historic opportunity to use the current phase of restructuring to put its economies on a sustainable path.

Many of the structural preconditions for sustainable development, such as participatory democracy and free markets, are now being introduced. What the area is still lacking is efficient technology and management practices to enable its industry to compete in world markets while ensuring environmental quality. One way to acquire such know-how is through joint ventures with Western companies.

In Poland, the basic framework for such joint ventures was established in December 1988 with the passage of the Foreign Investment Law. Asea

Brown Boveri (ABB) saw this as a signal to transform its existing business links with Poland into full-scale joint ventures. In a very short time, ABB was able to turn a fundamentally inefficient and poorly managed operation into a center of excellence for important products.

ABB was created in January 1988 with the merger of the Swiss Brown Boveri Company and the Swedish Asea group to form one of the world's largest electrical engineering companies. It has moved quickly to take advantage of the commercial opportunities generated by the failure of centrally planned economies: by 1991, the company had established approximately 20 joint ventures in Eastern Europe.

In Poland, ABB had two license agreements dating from the early 1970s. Zamech supplied a full range of steam turbines and Dolmel produced hydrogen-cooled generators. Both companies, however, lacked skills essential for survival in a competitive market economy. Furthermore, due to low efficiency and generally poor practices, levels of pollution and waste production were unacceptable. Prompted by the new foreign investment law and based on the company's good experience as a licensor, the management of Zamech approached ABB in December 1989 with a proposal to form a fully fledged joint venture.

By May 1990, ABB Zamech Ltd. had been established; ABB Dolmel followed four months later. According to Barbara Kux, president of ABB Power Ventures of Zurich, a new unit formed especially to manage the transformation of these companies, "for ABB, a joint venture of this type is intended to provide long-term access as an insider to a growing and potentially huge market for energy-related products. Naturally, ABB also hopes to take advantage of lower production costs." The benefits for Poland are also clear: "Through successful technology transfer, East European industry can become more efficient, producing quality products with less pollution and using less resources per unit of production," Kux notes.

Before these benefits could be realized, however, ABB, Zamech, and the Polish government had to go through a complicated process to establish the value of Zamech, based on its land, buildings, fixed assets, work in progress, contracts, and goodwill. Since at the time a foreign company was unable to own land in Poland, a new company had to be founded to own land and buildings. In addition, new license agreements and the future ownership structure had to be negotiated. Eventually, it was agreed that 76 percent of the new company should be held by ABB, 19 percent by the State Treasury, and 5 percent by ABB Zamech employees. The new management immediately embarked on a two-year turn-around program, focussing on basic restructuring, know-how transfer, and technology transfer.

ABB's first step was to reorganize each of the three business areas (foundry, turbines, and marine equipment) into independent profit centers, a task completed in just four weeks. For decades, Zamech had been a big overhead operation with unclear managerial authority and no true understanding of product-line profitabilities. ABB specialists in general management, production, finance, control, quality, and technology were designated as "restructuring agents" to work closely with a core group of "change agents" from the local management to identify priorities.

Eleven improvement projects in critical areas, such as reducing production time, improving quality, and reorganizing the sales function, were established. Every two months the status of each project was reviewed by a steering committee. Success was rapid: ABB's finance and control system was introduced within five weeks, following intensive training of local staff. The restructuring process was not without casualties, however, and after the initial reorganization measures, the number of employees had been reduced from 4,500 to 3,800. Further job losses could be needed to improve productivity. However, ABB believes that the new leaner company offers better long-term prospects for the remaining employees and good chances of employment growth in the future. In fact, ABB has already introduced two new business areas, district heating and power plant services, that should provide new jobs.

An extensive training program was established for both managers and staff to transfer the necessary know-how to achieve Western standards of production efficiency and product quality. A Total Quality Management program (TQM) has been introduced, and every employee of ABB Zamech will be given basic quality training, with a focus on quality awareness.

ABB was also concerned to build up the skills base of the new Polish management. One basic barrier was communication. After the merger, ABB Zamech adopted English as its common language. However, the Polish management spoke little or no English, and so a systematic program of language training had to be established. Thirty of Zamech's Polish managers were then provided with a brief, intensive Master of Business Administration program by the INSEAD business school outside Paris. The program contained elements of business policy, marketing, finance, manufacturing, and human resources management, as well as training sessions for specific functions, such as sales or purchasing.

The transfer of technology was achieved by confirming Zamech's existing license with ABB to produce steam turbines and signing new agreements for both steam and gas turbines. In addition, ABB Zamech has been chosen as a "center of excellence" for a specific gas turbine technology within the ABB group. The transfer efforts were so successful that the first turbine made in Poland left the factory after a record short production time of 10 months. Another new license between Georg Fischer and ABB Zamech for steel castings will dramatically improve the quality of ABB Zamech's castings. ABB Zamech has also established license agreements with Ullstein of Norway to produce marine propellers, with RENK of Germany to manufacture industrial gears, and with ABB's I.C. Möller of Denmark to produce district heating equipment.

After 18 months, the results of the restructuring program had already become evident. Through the use of Quality Action Teams, cycle times have been reduced by 30–50 percent in the turbine business area. For steam turbines, the current cycle times match ABB worldwide averages. Although no specific activities have yet focussed directly on environmental issues, Lennart Alm, quality manager for ABB's Polish ventures, stresses that the TQM activities have a positive environmental effect through increased awareness and efficiency. Efficiency programs such as concentrating production and using state-of-the-art production technology and closed-water cycles greatly reduce emissions and resource use. Further progress is achieved by eliminating the most polluting processes and maintaining equipment more carefully.

By 1991, ABB Zamech was using 32 percent less electricity, 29 percent less gas, and 34 percent less water per dollar unit of production compared with 1989. In addition, emissions contain 71 percent less dust, 50 percent less nitrogen oxides, and 39 percent less carbon monoxide. Past emissions records for sulfur dioxide are uncertain, but from 1991 to 1992 emissions are expected to fall by 92.5 percent. The reductions in pollution and resource use (electricity, water, and gas) are expected to continue.

Lessons Learned

• Companies in Eastern Europe can be made efficient, achieving productivity and quality as high as Western companies. This can be accompanied by rapid reductions in pollution and resource use toward Western levels.

• Success is often achieved through the total commitment of all parties involved and through a willingness to leave as much as possible in the hands of local management.

• Considerable moves toward sustainable development can happen very swiftly.

Contact Person

Barbara Kux
ABB Power Ventures Ltd.
P.O. Box 8131
CH-8050 Zurich, Switzerland

Case 12.3
Eternit: Technology Cooperation for a Safer Working Environment

After World War II, enterprises for the production of asbestos-cement products were founded in several Central American countries by the Swiss Eternit Group (SEG) together with local partners. During the 1970s, asbestos became the object of increasing criticism. Fibrosis was recognized as an occupational disease common among workers in insulation, construction, and other industries, and evidence was growing that asbestos fibers, when inhaled, increased the risk of lung cancer. In several countries, trade unions campaigned vigorously against the use of asbestos, and demands for a ban became louder and louder.

The group management of SEG thoroughly evaluated the social and political dynamic of the health risks controversy posed by asbestos. The conclusion was that sooner or later the opposition against this raw material would create serious problems, even though asbestos was being handled in increasingly safer ways. Since scientific opinion about a safe threshold of exposure to asbestos dust continued to differ widely, a strategy to continue producing and using the material under "safe" conditions could not be defined.

After carefully weighing risks and opportunities, SEG decided in 1980 to adopt the precautionary approach and to replace asbestos entirely by 1990 or earlier where feasible. This decision was taken without having a technical alternative ready, and many specialists in the field kept arguing that this was outright impossible. In fact, it proved to be a difficult undertaking.

Finding a substitute for the "miracle fiber" was not enough. No fiber with comparable characteristics but without the health risks involved was available. Fibers of different sorts had to be blended and used together with modified binders. This required fundamental changes in a well-established technology, including the preparation and mix of raw materials and the manufacturing technique as well as final treatment of the product. Leadership and tenacity of management would be required to motivate the research and development team to stick to the target. Moreover, solutions in different parts of the world would have to be tailored to local circumstances, climate, and building traditions.

SEG launched a campaign with its licensees to participate in the search for substitute raw materials and for new manufacturing technologies. The engineers and technicians working at SEG's Costa Rican subsidiary, RICALIT, argued that European asbestos substitutes were inappropriate for Latin American conditions, being too costly and creating balance-of-payments problems, because they would have to be imported. The company encouraged them to develop technologies adapted to local conditions. After several false starts, the first trials with reinforced cement, later called Plycem, began in 1981. The new material was made up of a blend of fibers essentially composed of different grades of cellulose, recycled newspaper, and used banana boxes widely available in this region.

Commercial production started in 1982 on an experimental basis using rudimentary machines. The first sheets of Plycem were dried in a simple baker's oven. The characteristics of the new material made it ideally suitable to be used as a wallboard, a market that could not be penetrated with asbestos-cement sheets. As this new market developed, it proved to have far bigger potential than that for roofing sheets. During the following years, close cooperation between the Swiss parent and Industries Eureka in El Salvador also resulted in the development of corrugated fiber-cement sheets for roofs in 1989. Following the pioneering work at RICALIT, SEG's other Central American subsidiaries are manufacturing similar fiber-cement sheets.

The technological challenge was not the only obstacle the company faced. At the start there was both inertia and negative reactions at different levels throughout the company. Some of the managers in Central America and in Switzerland were not convinced by the warnings of asbestos health risks. There were also economic concerns. The production line for Plycem proved to be more expensive than the existing asbestos equipment. RICALIT had to invest $3.7 million in addition to $1.5 million in research, development, and testing costs. And the manufacturing of the new fiber-cement product was more complicated than expected. The company's marketing people greeted the new material with reluctance. The risk of losing core traditional buyers seemed very real. Existing product lines had to be redesigned and a new marketing strategy had to be devised, all of which required money and time.

Despite the initial opposition both within and outside the enterprise, the fiber-cement sheets soon proved themselves in local markets. The advantages of the new material became widely recognized. RICALIT found it was able to offset higher initial investment costs by lower material costs, making Plycem actually cheaper to produce. The final product was both less expensive and more manageable than asbestos cement. The customers got a better and safer product—and it costs less. Architects and designers found more possibilities for their creativity; it was, as one of them remarked, "easier to play with the material." City planners were also able to develop a new type of inexpensive dwelling using Plycem.

In El Salvador, Industrias Eureka began to produce low-cost prefabricated houses made of reinforced cement sheets. For the governments of the region, the removal of asbestos brought significant economic benefits. In 1982, RICALIT spent $700,000 to import asbestos. The introduction of Plycem cut hard-currency foreign exchange requirements to zero, since all the raw materials are available locally.

For the corporation as a whole, there have been many other advantages. After an introductory "shakedown period," the discovery of the asbestos substitute brought a fresh drive into the whole business. The skills and motivation of employees were improved. Last but not least, the asbestos-free, fiber-cement sheets eliminated potential health risks, while the use of recycled material and of waste made it possible to minimize the consumption of natural resources.

Production of the new materials grew rapidly and became highly profitable. In the mid-1980s, RICALIT at times had difficulty keeping up with demand. Sales grew from just over $3 million in 1983 to more than $6 million in 1991; this compares with asbestos-cement sales of under $3.5 million in 1982. The company is currently investing an additional $1.5 million to increase capacity and keep up with rising demand. SEG sees considerable potential for marketing Plycem in other parts of the world with similar climatic, social, or economic conditions, especially in the rest of Latin America. The United States also represents a promising market, and Plycem has been approved for import by the U.S. authorities. Plycem could thus be one of the many examples of South to North technology transfer, earning valuable hard currency.

Lessons Learned

• Environmentally safe substitute products can often provide better services at lower costs than existing materials.

• Strong and determined leadership is needed to overcome resistance to change, particularly if the replaced material, like asbestos, is a commercial success.

• Technology cooperation offers the possibility of producing competitive products in developing countries that could also succeed in industrial-country markets.

Contact Person

Marvin Montenegro
General Manager
RICALIT S.A.
Apartado 3482
San José, Costa Rica

Case 12.4
S.C. Johnson: Catalyzing Improved Supplier Performance

Changing regulations and consumer expectations mean that companies will have to reexamine their supplier relationships if they are to build sustainable business ventures in the future. In many companies, a growing amount of product components come from outside. In addition to ensuring that these meet quality, cost, and time requirements, managers will increasingly have to check that they meet environmental standards.

A commitment to life-cycle product stewardship, however, goes beyond assessing the environmental impact of the component itself. It means checking that the production and sourcing processes for that component are environmentally sustainable. Although introducing environmental criteria into supplier contracts could be seen as an added burden, as with quality considerations, companies are finding that the environment can be a stimulus for unexpected improvements and stronger business relationships.

S.C. Johnson, a U.S. producer of household cleaning materials, had commercialized innovative environmentally responsible packaging across its consumer and commercial businesses in 1990, including introducing the "enviro-box," which reduces the amount of material used in packaging by 92 percent; increasing the recyclable content in packaging for its floor care products; developing new 100-percent recyclable PET bottles for the product Future; reducing the packaging content of another Johnson product, Shout, by 90 percent; and introducing 100-percent recyclable packaging for its line of carpet cleaning materials. The company believes that it achieved a competitive advantage through working with suppliers to be the first to introduce these environmentally improved package options into the marketplace.

In November 1990, S.C. Johnson adopted a series of worldwide, long-term environmental goals consistent with the concept of sustainable development. The company recognized that to achieve these it would need to foster relationships with suppliers on environmental issues and to improve environmental performance further. The goals included, for example, phasing out certain environmentally suspect chemicals, which inevitably meant telling suppliers to find acceptable replacements. "Clearly, the company's environmental messages could have been communicated by letter, phone, or small group meetings," says Jane M. Hutterly, director of environmental actions worldwide. "However, none would effectively convey the seriousness or immediacy of the commitment S.C. Johnson was soliciting on such a global basis." Instead, the company decided to host a high-level conference for its suppliers and managers to bring attention to the new environmental commitment and to catalyze action to meet the company's goals.

The conference was a natural extension of the company's highly successful Partners In Quality Program, established in the United States in 1988 to identify and recognize suppliers that demonstrate the ability and willingness to work with the company for continued agreed-to quality improvements of products and services. Seventy-three suppliers participate in the process, 23 of whom have been recognized with the Partners In Quality Award for sustained improvement in quality levels of incoming materials.

Ninety-five representatives from 57 of the company's top 70 supplier organizations worldwide, supporting one or more of its major businesses, attended the "Partners Working For A Better World" conference held in February 1991 at corporate headquarters in Racine, Wisconsin; 20 percent of those attending travelled from their world headquarters outside of the United States. S.C. Johnson employees constituted the balance of the 186-person audience, including the heads of the three major business units along with key members of their management teams from the research, development and engineering, marketing, sales, and manufacturing areas.

In the keynote address, S.C. Johnson president and chief executive officer (CEO) Richard M. Carpenter presented the company's long-term environmental goals and focussed on past S.C. Johnson environmental achievements that had resulted from successful partnership efforts between the company and its suppliers. He highlighted instances in which improved environmental performance had brought a bottom line gain as well, such as the company's move to water-based formulations from solvent-based ones. Carpenter called for the expansion of the existing partnerships into an active environmental commitment to achieve specific, measurable goals.

"I believe the 1990s will give birth to a broad-based technical revolution prompted by environmental needs, and will result in unprecedented new business development," Carpenter told the group. He then called

for suppliers to review their existing low-environmental-impact technology and look for new applications or modifications to meet the needs of
its customers first, rather than apply all their resources to developing
new technology. Further, suppliers were encouraged to review their
operations, on a self-monitoring basis, to address the expectations held
by the environmental, government, and international business communities and to share their findings with the company.

The conference included presentations by well-known representatives
of those communities. In addition, some of S.C. Johnson's own customers
explained their environmental expectations. This helped reinforce S.C.
Johnson's message of increasing environmental interdependence among
companies along the entire product chain. Following the presentations,
participants joined 1 of 10 working groups to identify broad action steps
to support mutually stated environmental goals. The discussions focussed on biodegradation, source reduction, recycling, and volatile
organic compounds. S.C. Johnson was careful to create a nonadversarial
atmosphere by treating all the audience the same way. It also made sure
that there was room for dialogue at the meeting, so that all participants
had the chance to express their opinions.

For S.C. Johnson, the conference was designed to explain that the
successful achievement of the goals could only come through working
with "those suppliers who could advance progress against our objectives
in the stated time frame," says Hutterly. The response from suppliers
both at the conference and afterward was overwhelmingly positive. As
one supplier wrote, "we left with a better understanding of what we, as
a supplier, must do to keep S.C. Johnson in the forefront of its various
markets and environmental issues." Suppliers provided literature on
their environmental performance and numerous requests were made for
ongoing communication to share new information. This is being managed by S.C. Johnson's research and development and purchasing departments.

In addition, a follow-up brochure, providing reprints of the speeches
and summary highlights of the working groups, has been sent to all
invited suppliers and company attendees. A newsletter has been established that will report successful partnerships related to progress against
the company's environmental goals. These mailings will reinforce the
company's commitment to sustainable development by providing real-
life examples to illustrate that industry can and is acting responsibly. A

number of S.C. Johnson subsidiaries are planning similar localized gatherings with their suppliers during 1992.

According to company representatives, one result of the conference has been to identify and accelerate the process of introducing further improved packaging projects, which will contribute to the company's three-to-five-year goals. These include a 50-percent reduction in the total volume of packaging material used by the company by the end of this decade and the elimination of specific chemicals that have potentially adverse environmental effects. A new refillable plastic container has already been introduced for the company's household products in the Australian and New Zealand markets. Further, the discussions at the conference have already led to packaging improvements in both the U.S. and Canadian markets.

In addition, S.C. Johnson and its suppliers believe that the conference will also have a ripple effect, stimulating improved environmental performance in other companies. According to a major packaging supplier, the conference "will definitely create a momentum that will not only change for the better business with S.C. Johnson, but doing business throughout a number of industries." The company is certain that the concept of supplier-customer conferences to promote sustainable development could be replicated with similar success by other companies.

Lessons Learned

• Customer-supplier partnerships are vital to achieve improved environmental performance.

• Conferences bringing together company executives, suppliers, and environmental experts are a useful mechanism for catalyzing joint action on environmental goals.

• Conferences can have an important ripple effect, helping to improve environmental performance throughout a number of industries.

Contact Person

Cynthia Georgeson
S.C. Johnson & Son, Inc.
1525 Howe Street
Racine WI 53403-5011 USA

Case 12.5
The Chemical Industry: Introducing Responsible Care

The chemical industry is perhaps one of the most environmentally controversial sectors. The process of making, distributing, and using chemicals can be dangerous, posing risks to workers and nearby communities. Many of the industry's products can also damage the environment if misused or released accidentally. Legislative action and public outrage has risen over the years through a combination of headline accidents in the industry, such as at Seveso, Bhopal, and Schweizerhalle, and smaller-scale, more local issues.

The industry's response to these problems has often been inadequate. As U.S. communications specialist Peter Sandman told the U.S. Chemical Manufacturers Association in 1990, "Your industry hasn't kept its emissions as low as you practically could. Perhaps more important, you haven't built a good record of open communications with the communities in which you operate. Too often you have been arrogant or uncaring; sometimes you have been dishonest."[1] The result, admits Robert Kennedy, CEO of Union Carbide, is that "people don't trust us."[2] To tackle the twin problems of poor performance and poor public regard, chemical industry associations worldwide have been adopting Responsible Care.

The idea was first developed by the Canadian Chemical Producers' Association (CCPA) during the 1980s, and has since spread throughout the world. Under the Responsible Care initiative, chemical companies are committed in all aspects of safety, health, and protection of the environment to seek continuous improvement in performance, to educate all staff, and to work with customers, transporters, suppliers, distributors, and communities regarding product use and overall operations. This commitment is described in a set of guiding principles.

The implementation of the Responsible Care initiative is managed by national chemical industry associations. Although the details vary from country to country, there are a number of common features:

• a formal commitment to a set of guiding principles on behalf of each company, for example by CEO signature;

• a series of codes, guidance notes, and checklists to help companies implement the commitment;

- the progressive development of indicators against which improvements in performance can be measured;

- an ongoing process of communications on health, safety, and environmental matters with interested parties outside industry;

- provision of forums in which companies can share views and exchange experiences on implementation of the commitment;

- adoption of a title and a logo that clearly identify national programs as being consistent with and part of the concept of Responsible Care; and

- consideration of how best to encourage all association member companies to commit to and participate in Responsible Care.

In the United States, for example, Responsible Care was introduced in 1988 by the Chemical Manufacturers Association (CMA). To maximize participation by the industry and to show strong commitment, CMA has made the implementation of Responsible Care an obligation of membership; all CMA members are thus participating in the initiative. The association has so far developed five codes of management practice on community awareness and emergency response, employee health and safety, distribution, pollution prevention, and process safety; a code on product stewardship is also nearly ready. The pollution prevention code requires member companies, among other things, to make an inventory of wastes and emissions and then to design a program in collaboration with employees and the community to reduce these at the source. Companies have to monitor their performance against the code; CMA will use these data to communicate industrywide results to the public.

Sharing information and successful techniques with other companies is an important part of Responsible Care. As Union Carbide's Kennedy comments, "most of the tools of good environmental management are free, neither patented nor proprietary." More than 1,000 executives and managers from CMA member companies have attended workshops on implementing the codes. Mutual assistance has proved to be considerable.

Through Responsible Care, chemical companies have a clear framework within which they can begin a dialogue with the public. The chemical industry believes that demonstrating both commitment and improved performance is the key to maintaining and retaining public acceptance of the industry. As Kennedy says to the public, "Track us, don't trust us." Both the CMA and the CCPA have established Public Advisory Panels, consisting of community activists involved in environmental and health issues, to facilitate this process. Pat Delbridge, a Canadian environmental expert, told an international seminar on Responsible Care in April 1991, "any industry program setting out to change corporate behavior must include the external perspective," adding that an "advisory panel keeps you honest and relevant and can give useful guidance."[3]

Responsible Care was not developed without overcoming several obstacles. Associations met with skepticism from both the community and also their own members. By working together with the skeptics, however, the associations have managed to win support and start the implementation process. In the United States, from the industry side H. Eugene McBrayer, president of the Exxon Chemical Company, believes that "Responsible Care is giving us a template for making real, lasting improvements," while Peter Berle, head of the National Audubon Society, has called the initiative "an important step in the right direction."

The commitment of the industry and the strength of the initiative have led to Responsible Care being adopted around the world. Since it began in Canada in 1985, it has been introduced in the United States (1988), the United Kingdom (1989), Australia (1989), France (1990), and Germany (1991). Belgium, Ireland, Italy, the Netherlands, Spain, and Sweden also now have programs, and chemical associations in other countries are expected to introduce Responsible Care in the near future, including a number in developing countries and the former Soviet bloc.

Lessons Learned

• Industry trade associations can and should play a central role in improving the environmental performance of their members.

• Programs developed in one nation can be used worldwide by national and regional trade associations.

• Companies within the same industrial sector can assist each other in improving environmental and safety performance.

Contact Persons

Geoff Chambers
Australian Chemical
  Industry Council
380 St. Kilda Road
GPO Box 1610 M
Melbourne, Victoria, Australia

Brian Wastle
Canadian Chemical
  Producers' Association
350 Sparks Street
Ottawa K1R 7S8, Canada

Louis Jourdan
European Chemical
  Industry Council
Av. E. Van Nieuwenhuyse, 4
B-1160 Brussels, Belgium

Lori Ramonas
Chemical Manufacturers
  Association
2501 M Street, N.W.
Washington DC 20037 USA

Koichi Nishikawa
Japan Chemical Industry Association
Tokyo Club Building
2-6, 3-chome
Kasumigaseki
Chiyoda-ku, Tokyo 100, Japan

Case 12.6
Leather Development Centre: Promoting Best Practice

Making leather has traditionally been a dirty and often unsociable exercise. Tanneries have been located outside towns, downwind and downstream to carry away noxious smells and wastes. Today, leather production still involves considerable use of water and generation of wastes. On average, one metric ton of raw hide yields only 200 kilograms of leather. The by-products include 50 cubic meters of wastewater, containing a range of chemicals, and half a metric ton each of wet sludge and solid wastes.

But leathermaking has the potential to play a more positive role in the transition to sustainable development. From a total life-cycle perspective, the leather industry processes a waste by-product—skins—from another sector (the meat industry) into a range of useful products. In fact, the leather industry can claim to be world's largest industrial sector based upon a by-product. Yet producing leather is a complex, multistage process. At each stage, wastes are created.

Increasingly, however, industrialists and researchers are finding ways both to reduce the generation of pollution and waste and to upgrade the efficiency of the production process. A wide range of techniques and approaches can be applied, from simple low-cost housekeeping improvements through the substitution of less damaging chemicals to the introduction of intrinsically cleaner production technologies. Innovative ways of using tanning wastes as raw materials for glue, fertilizer, and animal fodder are also being developed. Together such improvements hold out the prospect of at least halving pollution with only marginal investments.

The challenge of diffusing "best practice" has been increased by the relocation of leathermaking during the last 30 years from the industrial to the developing world, where pollution control regimes are often weaker and where resources for environmental protection are scarce. In many developing countries, the leather industry has played a central role during industrialization; it is agro-based, labor-intensive, and adaptable to small-scale, low-technology production. At the global level, the International Council of Tanners has recently adopted a environmental code of conduct, while the U.N. Environment Programme's Cleaner Production Programme has established a tanning working party, which has collected numerous success story cases.

Real change, however, occurs at the local level. There is no such thing as a standard tannery; processes differ widely depending on location, sophistication, and the market for the final product. International best practice has to be translated into the local context; pilot projects are often needed to demonstrate the feasibility of change. Research institutes and trade associations together play a vital role in this process. In Kenya, the lead agency for the tanning industry is the Leather Development Centre (LDC) at the Kenya Industrial Research and Development Institute (KIRDI).

A weak institutional and legislative framework has meant that Kenyan industry has to date not received a sufficient regulatory push for improved environmental performance. In addition, lack of information and financial resources has limited the ability of local companies to assess and install cleaner technologies from abroad. An expert working at the Leather Development Centre sums up the challenge thus: "It is common among Kenyan tanneries to rely on traditional methods of processing leather, resulting in heavy production and discharge of pollution in the wastewater."

Collective efforts led by research institutes and/or industry associations can help spread the costs of improvement while minimizing the inherent risks of innovation. This is the role that the Leather Development Centre plays. It has been developed in phases over 10 years with technical and financial assistance from the United Nations Industrial Development Organization (UNIDO) and the Federal Republic of Germany. The center's aim is to provide advice and support to local companies on two interconnected issues: identifying and developing enhanced process and product technologies, which will boost the industry's domestic and export potential, and developing and diffusing better ways of reducing the environmental hazards from leather production. The LDC has established a pilot tannery plant, a leather design and production unit, a quality control laboratory, and a wastewater demonstration unit at its facilities in Nairobi. Although most of its funding comes from domestic and external government sources, a nominal fee is charged for the center's services.

The first step toward sustainable development for the leather industry is the creation of awareness among industrialists of the need for change. Once awareness has been raised, the center can provide tools to help companies overcome pollution problems. It uses its pilot production and treatment plants as models for training and demonstration purposes during seminars and workshops. This is coupled with regular visits to the tanneries themselves by LDC experts, and targeted assistance programs.

The LDC also works closely with the Kenya Tanners Association (KTA), of which it has recently become a member. During a tour of LDC facilities in September 1991, a leading KTA representative said, "the research and common facilities provided by the LDC are of great help to our members who lack the necessary resources for this kind of innovation." The manager of the Alpharama Tanneries added that "the LDC's research work, guidance, and advisory services are indispensable."

The LDC has made progress in a number of areas of pollution prevention in recent years:

Hides and Skins Preservation. Traditionally, hides and skins have been preserved with salt. However, this salt has to be soaked off before the leather can be processed, leading to problems with saltwater effluents. To avoid the use of salt, the LDC has supported the use of air drying, which is now used on more than 95 percent of Kenyan hides.

Unhairing. Sodium sulfide and lime are normally applied to get rid of remaining hair, resulting in high levels of sulfide pollution in wastewater. An alternative promoted by the LDC is to paint the hides manually with a special solution, and then recover the hair for making brushes. But this is a time-consuming process, and is only used in Kenya's smaller tanneries.

Tanning. Chromium is the standard chemical for tanning; it is highly effective at producing quality leather products, but also leads to toxic wastes. The LDC has been investigating different process formulations that favor better chromium uptake in the leather. In addition, the center has made a major breakthrough in the development of a vegetable tanning agent. This mimosa-based compound is available from local plantations and is thus cheaper than chromium. It has been taken up by some small and medium-sized tanneries, which use the mimosa process to produce leather straps for the famous Kiondo handwoven sisal baskets, which earn much-needed foreign exchange for Kenya. A manager at the Ndakani tannery, which has switched to mimosa, says, "I am very happy with the results so far. Using mimosa has saved money and cut down on chromium pollution."

Finishing. The conventional method of finishing leather using solvent-based top coats results in emissions of volatile organic compounds. In some industrial countries, solvents have been banned and replaced with water-based top coats. The LDC has started to warn local companies of the dangers of these chemicals, and has carried out tests at its quality control lab to demonstrate that water-based solutions can provide equivalent product quality.

Recycling Wastes. As part of its drive to cut chromium wastes, the LDC developed a way of reusing chrome wastes to tan split leather. This technique is now being used at the Alpharama Tanneries. Following trials at the center in the early 1980s, the Limaru Bata Shoe company now uses its leather shavings to produce leatherboard for shoe soles. One of its directors explained the role played by the LDC: "Leather shavings used for making leather boards are taken to the LDC for quality testing because of the availability of testing equipment and skilled personnel for analysis and advice." He added that "the support from UNIDO for the

LDC has been very helpful, and means that the majority of tanneries are now able to perform almost to full capacity with less damage to the ecological base."

Currently, none of the 15 tanneries in Kenya have pollution monitoring systems, which severely constrains their ability to introduce effective pollution prevention options. The LDC uses its pilot production plant to test new ideas and approaches for improving productivity and the treatment of wastes. Samples are also collected from local tanneries and assessed for their pollution content. Experts from the center then give the results to the company concerned, and provide assistance where needed to help reduce pollution.

The success of the Leather Development Centre can be measured by the support it receives from both local industry and international organizations. The KTA, for example, is planning to jointly finance future training courses and research work on environmental management issues. Meanwhile, the quality of the monitoring and testing work undertaken at the facility has been a contributing factor in the grant of UNIDO funds to upgrade the pollution control equipment at the Sagana tannery.

Lessons Learned

• Collective industrial research facilities can play a vital role in promoting cleaner production through collaborative demonstration and pilot projects.

• Financial incentives are needed to help the spread of cleaner technologies. For many developing countries, this means refocussing bilateral and multilateral assistance and cooperating with international institutions.

• A clear institutional framework is needed from government to guide the management of industrial waste problems adequately.

Contact Person

Dickson Songok
Leather Development Centre
P.O. Box 30650
Nairobi, Kenya

13
Managing Stakeholder Partnerships

Sustainable development incorporates the notion of "extended stakeholders": broadening the constituency, in time and space, of those affected by industrial actions. Successful organizations will be those that are better than their competitors at "adding value" for their stakeholders—not only customers but also employees, investors, suppliers, and local communities. The challenge is to bring about a true partnership, one in which both partners benefit and both pool their skills and energies to improve the quality and sustainability of corporate activity and the environment.

Case 13.1 looks at how Northern Telecom has worked with the U.S. and Mexican governments to tackle an international environmental problem—the depletion of the ozone layer. Case 13.2 describes how the Wildlife Habitat Enhancement Council improved conditions for wildlife at the site of a major industrial development project. An innovative idea by the Swedish government encouraged industry, as detailed in case 13.3, to develop and market a new energy-efficient line of appliances. Case 13.4 looks at cooperation between government, industry, and researchers to enhance tropical reforestation.

Case 13.1
Northern Telecom/Mexico: Technology Cooperation to Halt Ozone Depletion

The 1987 Montreal Protocol on Substances that Deplete the Ozone Layer brought to the attention of the world that the destruction of the stratospheric ozone layer can be attributed to emissions of chlorofluorocarbons (CFCs) and halons. These chemicals are extremely stable and do not contribute to local environmental or health and safety problems in the workplace; hence they have been popular in industrial use for the last several decades. They do, however, migrate to the stratosphere, where they decompose when exposed to ultraviolet radiation, releasing chlorine and bromine. These chemicals interact in the ozone layer and each atom of pollution can destroy anywhere from 10,000 to 1 million ozone molecules. As ozone depletion can cause a worldwide increase in the incidence of skin cancer, among other problems, it is considered one of the most serious environmental problems confronting the world.

The Montreal protocol and its subsequent amendment in London in June 1990 proposed eliminating CFC and halon use by the end of this decade. This has been agreed to by 93 nations. Achieving this goal is an ambitious task that requires both government and industry cooperation. Since the protocol was signed, industry has started to eliminate the use of the chemicals, often ahead of government requirements.

One company, Northern Telecom, used CFC-113 solvents to clean electronic circuit boards in its global manufacturing operations. Beginning in 1988, it joined other electronics manufacturers in making technological advances in cleaning electronic circuitry without using CFC solvents. Northern Telecom successfully eliminated the use of all CFC-113 solvents in its worldwide operations—from 1 million kilograms of solvent use in 1988 to zero consumption by December 1991. It spent approximately $1 million, but realized savings in the vicinity of $4 million by not purchasing CFC-113 solvents, by not having to pay U.S. CFC taxes, and by reducing waste disposal costs.

Governments, too, have begun to use legislation and regulation to control the use of ozone-depleting solvents, and they have actively encouraged and supported industry in reaching the targets set by the Montreal protocol. Mexico is one of the countries that immediately embarked on an ambitious program to eliminate the use of these solvents throughout the country. The government understood that it would require industry cooperation to reach its targets. In particular, it needed technical assistance to identify and implement new cleaning processes for its growing electronics industry. A special partnership was formed in 1991 between Mexican and U.S. government agencies, the Industry Cooperative for Ozone Layer Protection (ICOLP), and Northern Telecom to create an innovative training and demonstration program to eliminate ozone-depleting solvent use in Mexico. (ICOLP is a Washington-based association of 17 multinational corporations augmented by governments and industry associations; it coordinates the open, worldwide exchange of nonproprietary information on alternatives to ozone-depleting solvents in the electronics industry.)

Mexico wishes to take a leadership position on this issue for several reasons. First, its interest in CFC reduction is well known, as a Mexican—Mario Molina—collaborated in the development of the original theory of ozone depletion in the 1970s. Second, Mexico is pursuing an ambitious economic program that includes the development of free trade relations with the United States and Canada. Mexicans will need to improve their environmental performance and image if they are to be competitive in a new North American Free Trade Zone. Third, as the first signatory to the Montreal protocol, Mexico was determined not only to meet the targets set, but also to exceed them. Its international credibility was at stake.

The electronics industry is one of the fastest growing sectors in the Mexican economy, and it is also the largest consumer of ozone-depleting solvents. The Secretaria de Desarrollo Urbano y Ecologia (SEDUE), the Mexican environmental agency, recognized that electronic companies needed assistance if the country was to achieve the targets set by the Montreal protocol. The whole industry required new technologies and techniques in order to reduce its dependence on ozone-depleting solvents.

In 1990, the U.S. Environmental Protection Agency (EPA) contacted Northern Telecom and several other electronics manufacturers. Under the auspices of the Montreal protocol, EPA was conducting a study in Mexico to determine the best way to organize Mexican industry to eliminate the use of ozone-depleting solvents. It was aware of the work done by Northern Telecom and needed technical experts to join teams of representatives from SEDUE, the Mexican Association of Industries (CANACINTRA), and the electronics and metalworking industries.

Northern Telecom agreed to provide technical assistance and cooperation for several reasons. Sharing such expertise and technical information conformed with Northern Telecom's commitment to environmental leadership. Moreover, the company had been selling products in the

Mexican marketplace for some time and was familiar with developments in the electronics industry. "The creation of an appropriate forum for the transfer of technology suited to the needs of local businesses is the ideal extension of Northern Telecom's activities in Mexico. Northern Telecom's involvement in this project simply makes good business sense," said Margaret Kerr, vice-president for environment, health, and safety at Northern Telecom.

Under the arrangement, Mexico agreed to finance a technical information office in Mexico City, travel and per diems for technical experts, technology demonstrations, and technical implementation—all with funds it has access to in the Interim Multilateral Fund of the Montreal protocol. (This is managed jointly by the World Bank and the U.N. Environment Programme; developing countries qualify for the funds if they have ratified the Montreal protocol and consume less than 0.3 kilograms of CFCs per person per year.) Northern Telecom and other ICOLP members provided the experts free of charge, and Northern Telecom agreed to manage the project.

Planning meetings took place between November 1990 and March 1991. The partners created an innovative and challenging training and demonstration program to eliminate ozone-depleting solvent use in Mexico. As a result, Mexico became the first developing country to pledge to phase out CFCs by the year 2000—10 years earlier than mandated by the protocol for developing countries. This puts it on the same timetable as industrial nations.

The program was launched at an introductory meeting in June 1991, and the first workshop for companies from Tijuana took place the following November. Participants heard from seven U.S. specialists on a variety of control and replacement technologies, including the use of water-based cleaning substances and "no-clean" options. The workshop focussed on practical problem-solving, and involved a dialogue between business practitioners from North and South on the best ways to move forward. For Andres Jimanez of Ensambles Magnetico S.A., which uses ozone-depleting solvents to clean magnetic heads for computer disks, the meeting "really opened my eyes" on the possibilities for replacing costly chemicals with water-based cleaners, saving money, and improving product quality. Jimanez stresses the importance of "talking face to face with experts who have already phased out these chemicals," adding that this means "we don't have to go through the same learning process, only adapt what others have done in our circumstances."

This case study shows how international cooperation between governments and business can be made to work, and how quickly. The Mexican program was conceived, developed, organized, and delivered by a small group of committed people in just nine months. All parties had a shared commitment to the elimination of ozone-depleting solvents. Northern Telecom and the other ICOLP members had already taken the risks and made the investments in alternative processes; the task for the Mexican technical assistance program was to build on these breakthroughs and broaden the range of companies that could benefit from them. Furthermore, the scope of the program was not limited to Mexico.

The aim is to use the Mexican experience as a basis for programs in other developing countries. A project to phase out the use of ozone-depleting solvents in the Thai electronics industry has now been agreed on, using the Mexican example as a model. In addition, one of the objectives of the Mexican program is training developing-country experts who can become the frontline team members in subsequent projects throughout Mexico and possibly in other developing countries party to the protocol.

Lessons Learned

• International cooperation between governments and industry from industrial and developing countries is a highly practical way of resolving shared environmental problems. Such cooperation expands the resource base of a group and gives it access to funds for projects more easily than would be the case if one party pursued the options independently.

• Personal interaction between industrial experts from various participating nations is the best way to transfer skills and technologies.

• Voluntary participation is a cost-effective way to stimulate change—it can save governments significant expenditures later on enforcement and monitoring programs while improving the environmental performance of industry throughout the country.

• Progress toward sustainable development is achievable at a faster pace than was previously believed possible.

Contact Persons

Margaret Kerr   Sergio Reyes Lujan
Vice-President, Environment,   Undersecretary for Ecology
  Health, and Safety   Secretaria de Desarrollo Urbano y
Northern Telecom   Ecologia
3 Robert Speck Parkway   Mexico City, Mexico
Mississauga, Ont. L4Z 3CH,
Canada

Case 13.2
The Wildlife Habitat Enhancement Council: Industry in Harmony with Nature

Animal and plant species are now becoming extinct at unprecedented rates. Although considerable attention is rightly focussed on the destruction of wildlife and ecosystems in the tropics through deforestation, other regions of the world are also fast losing their traditional range and depth of species. In Spain, for example, native forests had been drastically depleted through overharvesting and population pressures by the middle of this century. In the 1950s and 1960s, the Spanish government sponsored a program to reforest the country using fast-growing eucalyptus. But eucalyptus do not provide any food for local wildlife and, if not managed properly, can crowd out the remaining native trees, such as oak and chestnut, and drain the soil of its nutrients.

In the Asturias region of northern Spain, a proposed huge industrial development by Du Pont offered an opportunity to reverse the spiral of wildlife decline. Working in collaboration with the authorities, the local community, Spanish conservationists, and an innovative U.S.-based wildlife concern (the Wildlife Habitat Enhancement Council [WHEC]), this major chemical company has been able to design its newest industrial facility in such a way that biodiversity will increase.

Du Pont has made promoting wildlife in and around its facilities one of its key environmental commitments. (See case 11.3.) In 1988 it was one of the founding corporate members of WHEC, whose mission is to promote the use of the extensive areas of land under corporate control for the benefit of wildlife. WHEC director Joyce Kelly believes that "thoughtful and cooperative projects can transform these areas into wildlife havens, thus protecting a crucial link in the chain of global biodiversity."

Through WHEC, more than 66,000 hectares are now being managed for wildlife in the United States at 140 corporate locations. WHEC's projects are designed and implemented with the participation of employees, local conservation groups, and the community at large. In October 1991, WHEC was awarded a Presidential Citation as part of the First Annual White House Environment and Conservation Awards. WHEC is now expanding its activities to Europe, Australia, and Latin America.

In 1989, when Du Pont announced plans to establish a $1-billion multiproduct industrial facility in Asturias, WHEC was invited to develop a Wildlife Habitat Opportunities Report and to advise the company on how its new investment could be leveraged for the benefit of local species. The result was a comprehensive plan for the design and management of the facility, made with the participation of the local community. As Bill Walker, Du Pont's project director at Asturias, says, "Working with WHEC has helped to both mitigate some concern in the community prior to construction and demonstrate the compatibility of industry and the environment."

The Asturias facility is located on 320 hectares of land in a rolling valley. A number of production units will be built over the next 10 years along the valley floor, for the most part producing high-value fibers and raw materials for fibers. The first unit, manufacturing fire-resistant NOMAX material, will come on stream in January 1993. The areas between and around these production units have, however, been dedicated to wildlife, amounting to about half the total area of the facility.

At each of its wildlife project sites, WHEC attempts to link natural and human resources by encouraging maximum participation and involvement from both the management and employees of the company concerned, as well as the local community. At the Asturias facility, this meant working closely with faculty and students from the University of Oviedo, the local conservation group Associacion Asturiana de Amigos de la Naturalez (ANA), and the widely respected local conservationist Enrique Pasqual. These groups and individuals are collaborating in the development of a comprehensive management plan to enhance the area for wildlife, improve water quality, and preserve local history and culture. They will be joined by an employee wildlife committee once production at the site begins.

The project is designed in such a way that the partners can contribute what they each do best. WHEC provides continuing overall guidance in recommending appropriate projects and linkages between people and local wildlife; Du Pont offers its land, management commitment, and support based on a philosophy of stewardship; ANA represents and recruits local support for environmental protection; and the University of Oviedo conducts and applies the scientific research essential to the environmental plan. These different partners are united, however, in the common goal of improving habitats for existing species and encouraging the return of species once common in the area.

The first challenge for the project was to halt the clear-cutting of land being purchased from local farmers. The compensation schedule arranged by the Spanish government had allowed the farmers to remove any structures and trees before handing over the land for development. This process threatened to destroy the few remaining native trees and exposed erodible soil to wind and rain. The small stands of oak, birch, buckthorn, and chestnut, as well as shrubs such as the strawberry tree and bayleaf, provide food, cover, perching sites, and nesting/denning cavities for a range of native animal and bird species. A potential disaster was averted through an education campaign, informing the farmers of the environmental benefits of preserving the trees. A detailed inventory of the plant and animal species on the site is now being made.

Du Pont also adapted the layout of the facility to match the contours of the valley and the needs of the wildlife. Standard construction techniques would have meant grading the entire site to one level in a single operation. This would have forced all wildlife out simultaneously, putting considerable pressure on adjacent areas by creating greater competition for breeding areas and food sources. Instead of taking this approach, the Asturias complex has been designed as a series of self-contained production facilities at different sites in the valley. The preservation of hills and stream beds ensures minimum erosion, good absorption of water runoff, and the protection of plants and animals. These environmental plans mesh well with the industrial requirements of the facility. Each production unit within the compound is devoted to a single product, which the company believes is an efficient way to produce a range of products at a single site. In addition, instead of building a large concrete drainage ditch, Du Pont has identified and protected natural drainage processes, "serving our needs and saving us money," says Walker.

Environmental criteria have governed the construction methods as well as the site design. This means that the various production facilities are being built sequentially rather than simultaneously. Sequential construction allows land that will eventually be developed to benefit wildlife in the interim. Species displaced by construction are thereby assimilated better into the surrounding terrain. Within the construction area, valuable tree species, including mature olive, chestnut, and oak, are being relocated. Du Pont expects to have a 60-percent success rate with these trees. Native trees and shrubs in the 160 hectares set aside for wildlife are being protected, and two test plots of 350 native trees have been interplanted with eucalyptus to determine how best to restore native woodlands and wildlife. During 1991, 25 hectares were reforested with native saplings.

The project goes further than simple restoration. Studies are currently under way to develop a wetland zone in the region. As Spain is on a migratory flyway for birds returning south from Iceland, Greenland, and the Soviet Arctic, the development of a wetlands would provide a resting or wintering area for numerous species of waterfowl. Local schools and the community will be involved in projects such as the development of a nature trail and a nesting box program.

Finally, when the facility becomes operational, it will use state-of-the-art clean technology. An independent environmental impact assessment commissioned by Du Pont concluded that production at the facility would have "minimal impact and assumable risks." The company expects to devote 15 percent of its investments on the facility on pollution control measures. As a result, there will be no groundwater pollution, according to site director Walker. In addition, the NOMAX production process has been modified to eliminate the use of two potential carcinogens. The unit plans to manufacture a raw material for lycra production that will produce "zero emissions."

WHEC and Du Pont hope that the project will become a demonstration area for other companies in Spain and the rest of Europe. According to Du Pont chairman Ed Woolard, "the wildlife habitat program, through involvement with WHEC, has high value to Du Pont. It focusses on involving employees and the community in worthwhile environmental projects while demonstrating that we can achieve a balance between a healthy environment and a healthy business."

Lessons Learned

• The interests of industry and wildlife can be complementary through careful planning and the full participation of all stakeholders.

• Industrial development can offer opportunities for enhancing degraded land areas.

Contact Person

Joyce Kelly
Executive Director
Wildlife Habitat Enhancement Council
1010 Wayne Avenue, Suite 1240
Silver Spring MD 20910 USA

Case 13.3
Electrolux: Designing Energy-Efficient Products

Often government intervention is needed to give the correct market signals by providing extra information to consumers, changing market prices, or stimulating companies to be more innovative in their product design through competitions or award schemes. This was the strategy adopted in 1989 by the Swedish National Board for Industrial and Technical Development (Nutek) to promote the introduction of more-efficient refrigerators.

Nutek's mission is to promote development and commercialization of energy-efficient technologies. Despite an average reduction in electricity consumption by refrigerators of 95 percent over the last 30 years, these appliances still account for 30 percent of household electricity consumption in Sweden. To speed up the product development process, Nutek decided to organize a competition open to all foreign and domestic manufacturers to design a super-efficient fridge-freezer. That was not all. Nutek was also concerned to ensure that the winning model met consumer requirements and would have a ready market. As a result, representatives of major refrigerator purchasers were invited to help draw up the terms of the competition. This procurement group, consisting of public, cooperative, and commercial housing corporations and representing 25 percent of the market for fridge-freezers in Sweden, together with government officials and a consulting company in early 1990 decided on the following criteria:

Energy Efficiency. The units could not consume more than 1 kilowatt-hour (kWh) per liter (l) per year. This compared with an average energy consumption of 1.4 kWh/l for new units, and 2.0 kWh/l for units already installed.

Environmental Compatibility. The units had to meet high environmental standards, including the minimization of ozone-depleting chemicals in the insulation and cooling systems.

Consumer Information. Energy consumption information had to be displayed to enable the customer to make an easy comparison between different models at the retail outlet.

In return, the winner was guaranteed an order of at least 500 units from the procurement group. In addition, Nutek agreed to provide a "reward" of Skr500 ($80) per unit to the winning company if electrical consumption did not exceed 0.9 kWh/l per year and to subsidize the purchase of 500 units to the tune of Skr1,000 ($160) each. The timetable for the competition was very demanding.

A total of five companies entered the competition, two of which were Swedish. In December 1990 the winner was announced as Swedish-based Electrolux, one of the world's leading manufacturers of refrigerators, freezers, cookers, microwave ovens, dishwashers, and washing machines. About 85 percent of group sales are outside Sweden.

The winning model consumes only 0.79 kWh/l a year; the company also entered a super-efficient model that uses only 0.53 kWh. In addition, Electrolux reduced the amount of CFCs used in both models, thereby cutting their contribution to global warming and ozone depletion to one tenth. Electrolux also agreed to gradually introduce a label with product information on refrigerators and freezers produced in Sweden. Although the winning unit costs about 10–20 percent more than a standard Electrolux unit, the cost will be paid back in about four years due to the lower energy consumption. The payback period will be even shorter if Sweden's low electricity prices rise to a European average, as expected.

If all manufacturers were to reduce their power consumption to that of the winning product, at least 2 terawatt-hours per year would be saved in Sweden 10–15 years from now—an amount corresponding to almost half the yearly production of an average nuclear power plant. On October 9, 1991, the first 80 new fridge-freezers were installed in the city of Gävle. While the winner originally was guaranteed the sale of 500 units, the publicity that the competition received meant that Electrolux received orders for more than 10,000 units by November 1991.

Although Electrolux admits that it was technically possible to have designed the winning model without government intervention, Bo Kylin, marketing manager of major appliances at Electrolux AB, argues that "before the competition the market just wasn't there!" But it was not just the competition that stimulated the company to act. According to Tord Kyhlstedt, marketing director, the most important stimulus for going from prototype to product was the prospect of large orders, allowing the company to spread its development costs and thus reduce the sales price. Electrolux expects that the use of the products in new housing units will spill over to the still larger replacement market, where purchases are mainly made by individuals and families.

For consumers, Electrolux, and society as a whole, the most important result of the competition is the creation of a market for energy-efficient appliances. To meet the new demand, Electrolux is introducing a whole new line of "super-efficient" appliances as an alternative to its standard range. The line introduced in fall 1991 includes the award-winning fridge-freezer, two new refrigerators, and two new freezers. According to Electrolux, the refrigerators have also set new efficiency records. Further models are being developed. Electrolux is also planning to market these products outside Sweden. Demand is initially expected to be centered in the rest of Scandinavia, Switzerland, the Netherlands, and Germany, where buyers are environmentally mature and susceptible to total cost advantages. According to Kylin, early indications suggest that Electrolux super-efficient refrigerators could be responsible for up to 50 percent of the company's sales in Germany in 1992. Electrolux's total market share is expected to increase slightly as well.

The competition obviously helped change the attitude and plans of a large multinational company. Electrolux now realizes the importance of maintaining public credibility and of demonstrating good intentions, or as Leif Johansson, head of major appliances, puts it "environmental impact is assuming a growing part of the corporate image, because of the serious public interest in environmental issues." One sign of this new commitment is the company's program to phase out the use of ozone-depleting CFCs. Electrolux has developed technology to stop leaks of CFCs during repair work and has made it company policy not to use CFC-blown foam in its packaging. Electrolux is also testing a facility to recycle CFCs, as well as steel, copper, and aluminum, from discarded refrigerators. Inspired by the German automobile industry (see case 16.6), Electrolux is starting to prepare its products for easy "demanufacturing."

Lessons Learned

• Environmentally related product improvements can be accelerated through innovative government actions, which work with the grain of the market.

• If the right market conditions are created, industry can quickly become eco-efficient.

Contact Person

Bo Kylin
Marketing Manager
Electrolux Major Appliances
Luxbacken 1
S-105 45 Stockholm, Sweden

Case 13.4
Mitsubishi: Cooperation for Reforestation

For companies that are in any way involved in logging or related activities, the issue of deforestation requires a reexamination of their policies and practices. Indeed, a number of leading companies have failed to appreciate the growing gap between their performance in tropical rain forests and public expectations.

For the Mitsubishi Corporation, a major Japanese trading company, this problem was symbolized in August 1989 when its president, Shinroku Morohashi, was among eight global political and business leaders chosen by American environmentalists as men "who will decide the fate of your children" because of their alleged involvement in tropical deforestation. Advertisements accusing the eight appeared in the *New York Times* and other major newspapers, and environmental organizations began to campaign against Mitsubishi. "Needless to say, this gave us some concern," according to Kyosuke Mori, general manager of environmental affairs. "A reassessment of our policies was urgently needed," he adds.

Japan is the world's largest importer of tropical timber, 92 percent of which comes from Malaysia. Although Mitsubishi has logging operations in this country, the company is not ranked among the top 10 Japanese importers of tropical timber. It appears to have been chosen for attack because of name recognition. None of the larger Japanese timber importers were targeted by the American campaigners. The Mitsubishi brand name, however, appears on automobiles and consumer electronics products around the world, although they are produced by companies totally independent of Mitsubishi Corporation. In the past, Mitsubishi had supported and encouraged government reforestation programs, contributed capital and technical skills, and added value to the local economy through the production of plywood and other products. It had also ensured that it did not clear-cut tropical forests.

The launch of the environmentalists' campaign against Mitsubishi made the company realize that this approach was no longer sufficient. A study group was formed to rethink the company's approach to both the direct impact of its logging operations and the wider global implications.

The study group reached four key conclusions. First, based on years of experience in logging, "we realized that existing technical approaches to reforestation would need to be revamped" in response to new economic and ecological considerations. Second, Mitsubishi accepted that the threatened destruction of tropical forests had become a focus of public attention. Academics, environmental organizations, and interest groups have called attention to a problem for which there was little information on the causes and even less information regarding solutions. The third conclusion was that the visibility of Mitsubishi Corporation's logging operations was disproportionate to its size: "Attacks on Mitsubishi Corporation created problems for other Mitsubishi companies with no involvement in our logging operations," admits Mori. Finally, the study group highlighted the link between deforestation and the much wider problem of global warming. As the largest supplier of energy—in the form of natural gas and oil—to the Japanese economy, Mitsubishi Corporation needed to understand the global warming issue, and not just in terms of the role of deforestation.

In October 1989 a global environmental committee was established to recommend concrete measures for action. Based on its recommendations, a 10-person environmental affairs department—the first for any Japanese trading company—was established in April 1990. President Morohashi explains the reasoning behind the move: "Although we have addressed specific issues in the past, we foresaw the need for a more coordinated approach to developing solutions to environmental problems." The new department was given the mission of supervising environmental audits of each business area within the corporation and maintaining a dialogue with government agencies and environmental organizations. In addition, it was also provided with funds to support technological research beneficial to the environment.

The department immediately focussed on the situation in Malaysia. Initial research made it clear that the existing body of knowledge about regenerating tropical rain forests was limited. Conventional reforestation programs were also flawed from both an economic and an ecological perspective. Economically, these programs had the drawback of reducing the diversity of tree species, thereby undermining the basis for continued consumer demand for tropical timber. It had the added ecological problem of transforming diverse natural ecosystems into single-species plantations.

The company wanted to undertake a project that would add to the body of knowledge about reforestation. In its research, the company had come across the work of Akira Miyawaki, an internationally renowned vegetation ecologist who heads the Institute of Environmental Science and Technology at Yokohama National University. Miyawaki argued that forests could be regenerated if indigenous species were used. He had proved his theory in Japan by directing more than 200 successful native forest regeneration projects and had studied botany in Thailand and Indonesia in preparation. Mitsubishi staff proposed a four-year, ¥280 million ($2.3 million) experiment to create a rain forest ecosystem in the state of Sarawak, Malaysia. The project had the full support of the Sarawak government.

According to Mori, the initial proposal was greeted with skepticism by some of Mitsubishi's managers. "They noted that this approach had no commercial record in the tropics. They saw that it would be considerably more expensive than conventional methods of reforestation. And they questioned the use of three full-time technical experts and the valuable time of many other staff members that would be required to see the project to completion." However, these doubts were overcome with the realization that if the project succeeded, it would have applications throughout the tropics. Even if it failed, the knowledge gained in the process would help shape future experiments. So in early 1990, the company gave its go-ahead. The timing could not have been better, as it coincided with the Yokohama-based International Tropical Timber Organisation's call on Malaysia to reduce forest logging in Sarawak by 26 percent.

Mitsubishi Corporation applied its organizational skills to the task and arranged for Professor Miyawaki to work in cooperation with Malaysia's National University of Agriculture. The project, located at the Bintulu campus of the University, involves a plot of about 50 hectares. In November 1990, the project group began collecting seeds of the diptercarpaceae species, a family of trees native to the region. A nursery was constructed on the university campus to grow seedlings, and in March 1991 the group began planting saplings on land at the site. If the saplings take root within the next year or two, they will grow without requiring any care. It is quite possible that a "primeval" forest could exist on that land in 50–100 years. Mitsubishi Corporation has encouraged the exchange of information about the project, holding a tropical forest seminar in Malaysia in July 1991, another in Bonn, Germany, in autumn 1991, and scheduling a third in Sarawak for September 1992.

It is important to make a distinction between restoration of lands already deforested and the wise use and preservation or restoration of existing forests. This case focusses on the latter, while case 17.4 discusses the former.

Interestingly, the logging issue has stimulated changes in Mitsubishi thinking that will have benefits in other areas. The corporation has not only established its environmental affairs department, but, says Mori, "our managers now realize that discussion of policy alternatives should be long-term, and not be limited by current scientific knowledge." The company's response to the issue has also had payoffs in terms of public image: Mitsubishi Corporation was voted top of nonmanufacturers in a 1991 *Nikkei Business* magazine survey.[1]

Lessons Learned

• There are opportunities to form innovative collaborative arrangements for technical cooperation among government, academic institutions, and commercial enterprise on a bilateral basis.

• Truly creative alternatives to environmental problems will only come as companies are willing to lengthen the time horizon over which they can be solved.

Contact Person

Kyosuke Mori
General Manager
Environmental Affairs Department
Mitsubishi Corporation
6-3 Marunouchi 2-chome
Chiyoda-Ku
Tokyo, Japan

14   Managing Financial Partnerships

Moves toward sustainable development will require new financial partnerships, as well as changes in the ways projects are financed. Individuals, banks, companies, and governments need to make "smarter" investments that not only show immediate profits but that are sustainable and generate positive environmental results and profits into the future. This chapter includes five cases that are examples of new financial partnerships that promote sustainable development.

Case 14.1 looks at the establishment of a Nordic financial institution to provide seed capital for private-sector sustainable development projects in Eastern Europe.

Small businesses and the informal sector play a key role in providing livelihoods for the poor. Case 14.2 discusses how FUNDES, a nonprofit foundation established by business, is providing both financial services and management training to smaller entrepreneurs in six Latin American countries. In Chile, another nonprofit institution, described in case 14.3, has developed and assisted the start-up of new private-sector companies that have generated economic growth consistent with environmental goals.

Case 14.4 looks at how a private-sector financial lending company has established a new program to help itself and its clients identify and manage potential environmental risks. And case 14.5 focusses on the new environmental investment funds that are providing investors, both large and small, with opportunities to give priority to companies moving toward sustainable development.

Case 14.1
Nordic Environment Finance Corporation: Financing for Sustainable Development in Eastern Europe

Investing in the new markets of Eastern Europe poses many risks for companies. Banks are often unwilling to engage in medium- and high-risk projects, and there is also a shortage of venture capital. In response to a worldwide funding gap in this area, several development finance institutions have emerged to support economic development, particularly through private-sector investments. Among these are the International Finance Corporation and the newly formed European Bank for Reconstruction and Development—the first international bank with a commitment to sustainable development written into its constitution. Nevertheless, both foreign donors and local recipients agreed that the pressing environmental needs of Eastern Europe required a specifically environmental development finance institution, and so in October 1990 five Nordic countries—Denmark, Finland, Iceland, Norway, and Sweden—established the Nordic Environment Finance Corporation (NEFCO).

NEFCO's mission stresses the importance of developing a thriving manufacturing and service sector to deliver environmentally enhanced products for Eastern Europe. Its role is to provide risk capital to help this process, by supporting long-term cooperation between Nordic and East European enterprises. Operations are limited to Czechoslovakia, Estonia, Hungary, Latvia, Lithuania, Poland, and the western states of the former Soviet Union. Furthermore, at least one Nordic company, and normally also a local company, must participate as active partners in the project. If these conditions are met, firms from other countries can participate as well. "It is the hope of the Nordic governments that these investments will stimulate local private enterprise and ultimately improve the environment in the recipient countries and indirectly also in the Nordic countries," explains a NEFCO official.

The board of NEFCO is composed of one representative from each member country, generally selected by that country's Ministry of the Environment. Day-to-day operations are handled by the Nordic Investment Bank, headquartered in Helsinki, Finland. The initial capital of NEFCO, contributed by the member countries, is approximately $50 million. Because of relatively limited funds, it focusses on small and medium-sized ventures. NEFCO's project appraisal procedures stress the importance of achieving environmental as well as economic benefits. Special priority is given to projects that have a regional environmental effect benefitting the Nordic region as well (for example, through a reduction of transboundary pollution). Projects could thus include production of equipment for wastewater treatment, which would reduce emissions into the Baltic Sea, and flue gas scrubbers to cut acidic emissions. NEFCO also considers projects dealing with the environmental modernization of energy and industrial facilities, as well as the establishment of environmental consultancy businesses.

Although the purpose of NEFCO is to assist projects that may be too risky to win private financial backing, all its projects must demonstrate a potential for generating profits over the long term. NEFCO aims to act as a catalyst to bring together possible business partners for long-term cooperation, preferably in the form of equity joint ventures. Its investments are primarily made in the form of equity participation, typically in the range of 20–35 percent of the total equity investment. In addition to providing equity capital, NEFCO can also make loans, on commercial-term market rates.

By the end of 1991, NEFCO had approved investments in six joint ventures—four in Poland, one in Russia, and one in Czechoslovakia. Three of the Polish projects are for the establishment of forestry, energy efficiency, and desulfurization consultancies. The Russian venture involves an engineering business to install desulfurization equipment for coal-burning power plants, while the Czechoslovakian venture is a public/private partnership in the waste management business. The other Polish project—NEFCO's first—involves the production of water treatment chemicals out of industrial waste at the Kemipol joint venture, near Szczecin. NEFCO's involvement is based on the serious damage that effluent from Polish rivers is causing the Baltic and the damage done to local health. "What Poland urgently needs is a combination of strong efforts to reduce emissions at source and an increased wastewater treatment capacity," says Harro Pitkänen, Director of NEFCO. But Poland also needs access to effective treatment technology and chemicals at a reasonable price.

During the summer of 1990, Kemira Kemi AB of Helsingborg, Sweden, had started assessing a project to manufacture water treatment chemicals in Poland. The company already had a number of production facilities across Western Europe for the manufacture of ferric sulfate, a commonly used chemical for sewage treatment. Ferric sulfate is particularly effective at removing phosphorus and reducing both biological oxygen demand and suspended solids. When used in connection with existing biological treatment processes, capacity can be increased with only marginal investment. Alternatively, new biological plants can be built more cheaply. However, local production of ferric sulfate was necessary to lower costs.

A solution emerged that helped solve a mounting hazardous waste problem at a Polish chemical company, the Zaklady Chemiczne Police (Police chemical works). Police is a large producer of fertilizer and chemicals, including titanium white. The production of titanium dioxide resulted in the generation of ferro sulfate waste; in the past, Police has stockpiled its waste, which has been associated with environmental hazards due to leaching of heavy metals.

NEFCO saw an opportunity to help establish a much-needed water treatment business and at the same time cut the company's waste problem. Following negotiations, the three partners—NEFCO, Kemira, and Police—established a limited liability company, Kemipol Sp. z.o.o., to plan, build, and run a plant to produce water treatment chemicals. The joint venture is owned 20 percent by NEFCO and 40 percent each by Kemira Kemi and Police. The plant will be erected on a piece of land adjacent to the Police works. This gives the facility an important cost advantage, because of the availability of the basic raw material—ferro sulfate—on site.

The construction of the new plant means that most of Police's waste stream can be used; Kemipol also intends to make use of the ferro sulfate stockpiles, bringing further environmental benefits. The other main raw material, sulfuric acid, is also available from Police. The plant began production in March 1992. A division of labor has been arranged so that Police will be the main supplier of raw materials and intermediate products, as well as electricity and heat, while Kemira will provide process know-how along with marketing and management support. The process equipment will be imported.

Kemipol is expected to be profitable within a short space of time, based on domestic sales as well as exports, bringing important foreign currency to Poland. As one Kemira executive says, "We have had only good experiences with NEFCO. Normally, we prefer complete control of a company but the Polish solution is working fine." The official status of NEFCO has also had the advantage of opening doors. It is hoped that NEFCO's presence on the board will add value as well in the future as the institution's experience of similar ventures grows.

Lessons Learned

• Environmentally conscious development finance institutions can act as important catalysts for the promotion of sustainable business ventures.

• Relatively small sums of public money can achieve high environmental and economic leverage through joint venture, equity-based investments.

Contact Person

Harro Pitkänen
Director, Nordic Environment Finance Corporation
Nordic Investment Bank
PB 249
SF-00171 Helsinki, Finland

Case 14.2
FUNDES: Promoting Small Businesses in Latin America

For decades, people in Latin America thought that the government had to be the main promoter of economic development, even though public sectors proved unable to meet those expectations. It has now become clear that long-term, more stable, indeed sustainable economic development must rely in the first place on individual initiative at the local level. Private entrepreneurship has been rediscovered as a crucial factor for economic and social development. In successful economies, small and medium-sized businesses are a core element—in some sectors accounting for the largest number of enterprises, the most people employed, and the most production.

In developing countries, people running small businesses usually have to face a range of restrictions, both internal and external, that often hamper their growth and development prospects. Credits, new technology, or know-how often are not available. Poor infrastructure and public services curb business activity: red tape and administrative requirements for the establishment and operation of small enterprises tend to push them into the so-called informal sector.

In 1985, a group of private Swiss business leaders established FUNDES, a foundation with the objective of promoting entrepreneurship in Latin American countries. The pragmatic starting point of the program is the direct involvement with small entrepreneurs looking for help in obtaining access to commercial credit. FUNDES, working with local foundations in various countries, aims to improve the competitiveness of this sector. The standard program offers:

• help in drafting a realistic investment project;

• assistance in completing a credit application to commercial banks that have cooperative agreements with FUNDES;

• subsidiary credit guarantees for small enterprises lacking sufficient collateral;

• credit monitoring with the lending bank;

• advice on organization, administration, finance, cost and personnel management, production, environmental issues, and marketing; and

• other specialized services such as accounting, factoring, and so on.

Even the best promotion program has little impact in an unstable and nonsupportive business environment, however. Small entrepreneurs, with their limited lobbying power, are all too often the victims of bureaucratic whims. Apart from governments' economic stabilization efforts, public-sector regulations for this group and their application have to be simplified. In the present situation, far-reaching institutional reforms are required to free private initiative. In six countries, FUNDES, with government and international institutions' support, has analyzed the business environment, which has led to deregulation proposals for small enterprises in several countries.

A number of principles guide FUNDES: The small entrepreneur is the agent par excellence of local private initiative. Widening the field of opportunities for such individuals is crucial. And programs should function in a decentralized system of local organizations, jointly sponsored by successful local business leaders who take the lead in establishing and developing them.

So far, the foundation is active in Bolivia, Chile, Colombia, Costa Rica, Guatemala, and Panama. FUNDES Switzerland provides the expertise needed to set up the service program and makes financial commitments in the form of loan guarantees up to a maximum of 50 percent of the loan.

By 1991, after operating for five years, the FUNDES network provided access to bank loans for 750 small entrepreneurs, involving a total amount of $7 million. The overall default rate to date is below 3 percent and the loss rate is under 0.3 percent of the total amount of credits guaranteed. During 1991, some 800 small businesses participated in FUNDES business administration courses, and the network provided consultancy services to 940 enterprises in areas such as investment planning, organization, cost calculation, production, and marketing.

One person helped by the program is Oscar García, who built up his "galvanizing" firm in Guatemala during the 1980s. At one point in 1990, a further expansion of his product range would have come to a halt due to lack of access to bank credit. He was unable to comply with the requirements of a formal credit application since his bookkeeping was incomplete; he did not know how to present convincing sales projections, and his assessed property value fell short of the required guarantees. He contacted FUNDES Guatemala, which investigated the enterprise and confirmed the weak points of the firm. In addition, the foundation identified a problem of contaminated wastewater.

Since the company's basic business fundamentals were found to be solid and growth prospects encouraging, FUNDES suggested that García participate in its vocational training courses. With the foundation's help he then worked out a complete credit application and the foundation provided him with a credit guarantee covering half of the $11,000 he had requested from a commercial bank. Given the relationship of trust between the bank and FUNDES, as a consequence of prior cooperation and a previously established agreement, the bank granted the loan. The monitoring was carried out jointly by FUNDES and the bank. García uses various FUNDES services, such as the accounting service, for which he pays a fee. He has since been able to create 13 additional jobs, to double the use of his installed capacity, and to increase his sales by 75 percent. As part of the process, García was able to reduce the water pollution from his operations. The foundation now requires its clients to address obvious environmental problems as a precondition for financial assistance.

Experience taught FUNDES to differentiate clearly between target groups—mini-, micro-, and small enterprises—since they each require different services. The foundation is also managing a micro-enterprises program in Panama providing average loans of $2,500.

In the area of mini- and micro-enterprises in Latin America, other institutions and networks—such as the Fundación Carvajal in Cali, Colombia, or the U.S.-based Accion International—have been successful in providing access to knowledge and small amounts of credit. The local partner of one of the most successful Accion programs in Latin America recently turned its credit program into a full-fledged commercial bank. Since 1987, PRODEM in Bolivia, a private voluntary organization under the leadership of local business people, has extended 75,000 loans of an average amount of $275, three quarters of them to women. Collectibles amount to only 0.24 percent of all loans. This program has transferred its financial activity to Banco Solidarion (BANCO SOL) with a view to increasing the program's sustainability and the capacity to fulfill increased demands for such services.

FUNDES, BANCO SOL, and similar initiatives take sustainability seriously. This implies the application of demanding criteria; if they are not followed, sustainability remains an empty concept.

Lessons Learned

• If the concept of sustainable development is to work, local entrepreneurs must have access to markets, credit, and know-how. Private ownership is good for sustainable development.

• Private initiatives can contribute efficiently to creating opportunities for small entrepreneurs in developing countries. People active in local businesses have to play leadership roles.

Contact Person

Ernst A. Brugger
FUNDES
Haus Inseli
CH-8867 Niederurnen, Switzerland

Case 14.3
Fundación Chile: Financing Technology Cooperation

Developing countries need new industries to generate the wealth to meet their needs and aspirations. But building the necessary entrepreneurial skills and technological capacity is a complex process requiring specialized assistance. Fundación Chile is a unique organization established in 1976 to foster the development of private enterprise in Chile. After an initial endowment of $50 million donated in equal parts by the founders—the Government of Chile and ITT Corporation of the United States—the foundation has relied wholly on income from its activities since 1986. These include the sale of technological services, income from technology-based demonstration enterprises, and earnings from financial investments. To identify new technology transfer opportunities, it relies on its own specialists and an international market and technology information network of more than 800 consultants.

The foundation's technological services include the development of production processes and products, and the preparation of projects. It holds courses and seminars in agribusiness, marine resources, and forestry, and publishes three highly informative trade magazines— *Informativo Agroeconómico* (Agro-economic Bulletin), *Aqua Noticias Internacional* (Aquaculture News International), and *Lignum: Bosque, Madera y Technologia* (Lignum: Forestry and Forest Products Technology). The foundation also creates pilot companies that demonstrate the feasibility of new technologies.

The foundation has created more than 30 pioneering enterprises, more than 20 of which were established in the last three years. Five companies have already been sold to the private sector, and the remainder are to be transferred to the private sector in the years ahead. Income received from the sale of these businesses is used to create new ventures. Salmones Antártica, Fundación Chile's pioneering salmon farming enterprise, which was vital for the success of the institution's salmon cultivation program, best illustrates the effectiveness of the demonstration enterprise model as a means for transferring technology.

During its early years, Fundación Chile began analyzing the potential for developing the nation's embryonic salmon industry. This research phase culminated in January 1982 with the acquisition of the open-sea ranching Pacific salmon smolt production facilities owned by Pasquera

Domsea Chile, a subsidiary of Campbell Soup, in order to establish Salmones Antártica. Later, encouraged by its own projections for the salmon industry, the foundation expanded Salmones Antártica's activities to include net-pen farming of Pacific salmon.

During the next few years the company built freshwater fish farming centers, seawater grow-out facilities, dry and wet fish feed plants, and processing installations, enabling it to produce smolts, salmon ova, and feed to satisfy its own and third-party needs, as well as fresh and frozen salmon for export.

By 1985, Salmones Antártica was raising most of its salmon in confinement, allowing its production to reach unprecedented levels in Chile. In addition, the company had developed the capacity to produce and mobilize all the raw materials required for its operations and resolve the sanitation problems associated with the rapid expansion of its farms. Underscoring the significance of these feats, during the 1987/88 season, Salmones Antártica became the first salmon farm in Chile to produce 1,000 tons of high-quality fish, triple the previous season's amount.

While still a foundation subsidiary, Salmones Antártica became the nation's largest salmon enterprise, capable of producing 1,500 tons of salmon, 5 million smolts, and more than 10,000 tons of salmon feed annually. Through its production of smolts and fish feed, the company made the development of other salmon ventures possible, and consequently accelerated the industry's growth.

To facilitate Salmones Antártica's continued growth, as well as the creation of similar companies—one of the main objectives of the demonstration enterprise technology transfer method—Fundación Chile provided technical assistance to and prepared projects for numerous sector entrepreneurs, thus helping them to participate in what has become one of the nation's most important export industries. Since the establishment of Salmones Antártica in 1982, more than 100 salmon companies have been formed in southern Chile. These companies, which collectively registered more than $110 million in export sales during 1990, have made significant contributions to the economic growth and prosperity of some of the region's most important areas.

In December 1988, after having completed its development cycle and fulfilled its technology transfer mission, Salmones Antártica was sold through an international bid to private-sector investors. The new owners have greatly expanded the activities of the company—still the unrivalled leader of the industry whose growth it helped to accelerate. In effect, Fundación Chile's salmon farming program, of which the Salmones Antártica project was the most important technology transfer instrument, is estimated to have boosted the industry's development during the last three years, resulting in important economic and social benefits for the country through the anticipation of profits to the entrepreneurs participating directly in the sector and those in related industries.

Other foundation-created companies transferred to private investors since the sale of Salmones Antártica include Procarne, which introduced boxed beef technology into the country; Berries La Unión S.A. and Berries La Unión C.P.A., which pioneered the production and export of the nation's fresh and "instant quick freeze" raspberries and blackberries; and Finamar, the country's first company to produce a smoked salmon for export.

The foundation's success was the result of a fruitful combination of Chile's favorable economic environment and the foundation's own unique, internally developed technology transfer method of creating pilot enterprises. This significantly boosted the foundation's credibility in the business community by providing tangible proof that it was willing to assume risks and capable of overcoming them.

Lessons Learned

• Nonprofit foundations can facilitate technology cooperation through the establishment of demonstration companies that are sold to the private sector when technical and economic feasibility is proved.

• Agriculture and fish farming can provide sustainable and profitable business development opportunities in developing countries when the government creates conditions favorable to free markets.

Contact Person

Patricio Barros
Fundación Chile
Av. Parque Antonio Rabat, sur 6165
Casilla Postal 773
Santiago, Chile

Case 14.4
GE Capital: Lending and Environmental Risk

Increasingly, sometimes through traumatic experiences, business is becoming aware that it may be exposed to environmental risks that could impose significant liability and costs. Burns Phillips & Co. of Australia had such a bitter experience: In 1987 the company bought two pharmaceutical plants in Italy. Subsequently it was saddled with environmental cleanup costs for pre-1987 problems. The remedial action cost more than the entire purchase price of the facilities.

The high potential of environmental risks creates problems for commercial lenders as well, since the borrowers may be unable to repay loans and, in some circumstances, the liability could extend to the lender. Environmental risks are fundamentally different from traditional insurable risks. As a result, they are often uninsurable for three different reasons.

First, for "normal" insurance to work, the individual incident to be covered must be limited to local damages only, in both space and time. It is no coincidence that the nuclear power industry in the United States sought to limit its liability to a fixed amount. The rest of the liability, should a disaster of unforeseen catastrophic levels occur, is to be borne by the government. Since environmental risks have a potential for almost unlimited liability, insurance is currently almost exclusively available through Lloyd's of London, and increasingly for limited circumstances of little use to lending institutions. Unlike the U.S. nuclear industry's situation, there is no statutory or financial limit on potential environmental liability.

The second reason for uninsurability is the likelihood of all policyholders being "hit" at the same time. An essential condition of insurability, namely that a pool of the insured lowers the risk that any one individual will make a claim, may not hold. For example, if a substance previously regarded as safe should suddenly be determined as toxic, liability could extend to each and every company manufacturing or using that substance. It is as though a single accident by one insured motorist could trigger claims for insurance coverage from every policyholder.

Third, to make insurance a viable business for the insurer, there need to be many small and well-defined incidents rather than a few unique or catastrophic ones in order to establish proper actuarial tables.

Thus it is not surprising that comprehensive environmental risk coverage is difficult to find. Companies are faced with theoretical liabilities hard to evaluate, but with the potential to bankrupt them—and increasingly there is no conventional insurance for these risks.

In ownership transfers, the first line of defense for a buyer is to seek warranties from the seller indemnifying the purchaser for any unknown and yet-to-be discovered environmental liability. Even if the seller is willing to provide such an open-ended warranty, it is only useful when backed by very well capitalized companies. In most private-sector transactions, this guarantee is not available. Since few such transactions are made in cash, lenders need to begin to have the ability to assess this contingent liability. Obviously the lender's primary concern is that the borrower, due to environmental liabilities, would be unable to repay the loan. A second concern is that the lender might be required to assume some portion of the responsibility for the remediation, cleanup, or personal liability suits flowing from the environmental risk.

GE Capital, formerly GE Credit Corporation, is a major commercial lender whose main activities are in the United States, with the remainder split between Canada and Europe. During the past several years the company has been operating very successfully. Pretax income has risen from $1 billion in 1988 to $1.4 billion in 1990, for an annualized growth rate of 25 percent. Company reserves of $1.4 billion are very solid. Yet the managers of GE Capital were increasingly aware that they, even as lenders, could be exposed to huge potential risks. In view of this potential exposure, GE Capital determined that they needed a systematic way, first, to obtain information about the risks of their clients and, second, to work with their borrowers to limit risks by preventative actions. A voluntary remediation plan is invariably less expensive than a court or government-mandated solution. The company hired John Campbell from the U.S. Environmental Protection Agency to help them design a program that would help them and their clients assess and minimize risks.

John Campbell, now vice president of environmental affairs, and his staff are responsible for comprehensive assessments of all environment risks associated with proposed lending transactions, with the objective of minimizing future losses of their borrowers from environmental problems. In the absence of precise legal, scientific, or operational standards, this is a complex task requiring innovative approaches. Most GE environmental affairs staff work is proactive—that is, staff try to identify areas of potential environmental risk before they assume crisis proportions. They have developed a standard environmental compliance service (ECS) document that must be completed as part of all loan applications. This 17-page checklist helps both GE and their loan applicants conduct an on-site inspection and assessment.

Often a great deal of the work is similar to a basic detective investigation. GE staff work with borrowing company personnel in going through the ECS check list. Who owned the property previously? What kind of commercial activity took place on the property and adjacent land (for both authorized and unauthorized activity)? Is it possible to talk with former employees who might know about unreported or unrecorded events or activities? Are any of the materials used on the land or in buildings now considered toxic or on any government "suspect" lists? Have the soils and the surface and underground water sources on the property been checked for suspected hazardous materials? Has any test drilling been conducted to determine if anything unusual is buried on the property? Has the current economic activity always been operated in conformance with all health, safety, and environmental rules and regulations? Does the company have an environmental auditing program to ensure that its future activities comply with all standards? As one GE investigator notes, "a lot of the ECS checklist is just common sense, but you would be surprised how often the companies had never asked themselves basic questions."

This program encourages borrowing companies to use environmental assessment and audits to help their own management to move toward sustainable development by preventing environmental damage and by acting early to address problems that have already occurred. Those companies will be well positioned to cooperate with their creditors in order to obtain a fair value statement of their company. Ultimately, companies that manage their environmental programs well will improve their credit ratings and become more attractive to lenders and investors.

Lessons Learned

• Lending institutions can protect themselves and their borrowing clients through careful assessment of existing and potential environmental risks.

• A competent environmental staff can identify potential environmental risks and propose preventative actions that will save money in the long run.

Contact Person

John Campbell
Vice President
GE Capital Corporation
260 Long Ridge Road
Stamford CT 06927 USA

Case 14.5
Jupiter Tyndall: Investing in the Environment

The role of the financial community is being closely examined for its impact on sustainable development. Sustainability requires environmental considerations to be factored into investment decisions for enterprises at all stages of development, from investment in new projects and technologies through to the upgrading of existing businesses. Currently this is happening only to a very limited extent because investors and their advisers do not have enough information on companies' environmental behavior.

Historically, where matters of environment and development are involved, concern about the role of finance has centered almost entirely on government loans and the activities of lending agencies such as the World Bank. Attention is now turning to the role of the private sector, such as commercial lending institutions and investors in companies quoted on international stock markets.

Investors in international stock markets are largely concerned with a company's ability to generate profits and pay dividends even though its products and production methods might have adverse effects on the environment. As a result, conventional investment analysis is based principally on historical or current income flows, and capital markets focus on short-term returns. Although the upgrading of products and processes, whether voluntary or enforced, may reduce short-term profits and dividends, the company will be creating competitive advantage for the medium and long term.

As stock markets have evolved, shares have become concentrated into larger holdings among fewer large institutional investors. In the United Kingdom, for example, institutional shareholders (pension funds, insurance companies, and investment and unit trusts) typically control more than 75 percent of the shares of listed companies. This situation can be turned to the advantage of sustainable development if these investors recognize environmental responsibility as an essential policy of companies in which they invest.

Jupiter Tyndall Group plc is a medium-sized, publicly listed U.K. investment management and banking group with funds under management and banking deposits totalling approximately $2.4 billion. This includes more than $84 million invested according to environmental criteria. Before the 1991 merger with Tyndall, Jupiter Tarbutt Merlin (JTM), now a wholly owned subsidiary, had been working for at least three years in environmentally responsible investment. Its managers had recognized that the response of the financial community to the environmental agenda was inadequate. The company realized that in many conventional investment vehicles, members of the public were contributing indirectly (and often unwittingly) to environmental degradation by investing in polluting and wasteful companies. JTM saw the "green investor" as the natural extension of the green consumer who emerged in the late 1980s, and viewed the need for a real investment alternative as a product development opportunity.

In April 1988, JTM launched the first specialist environmental fund, the Merlin Ecology Fund, an authorized unit trust. Its objective is to invest in companies making a positive contribution to protection of the environment and wise use of natural resources. Although three years is too short to assess performance, in the second half of 1991 the fund was standing fifteenth out of 161 funds (Micropal statistics) in the international growth sector.

The launch of the fund was inspired by the development of similar funds in the United States and was followed by increased interest among other investment groups there and in the United Kingdom. A growing number of environmental funds are operating in North America, Western Europe, and Australia.

At the end of 1991, less than four years after JTM started the Merlin Ecology Fund, more than $468 million is invested in environmental funds in the United Kingdom, demonstrating an interest among investors in acquiring shares in environmentally responsible companies.

In addition to individual investors, the pension funds of local authorities, trade unions, educational and religious organizations, and charities have been among the first to respond to this new investment agenda. Their support has spawned initiatives such as the Environmental Investor's Code.

Prior to the opening its Ecology Fund, JTM established the Merlin Research Unit, realizing that it would have to provide investors with an understanding of the environmental policies being followed by a company, and therefore a fuller understanding of the value of their investments. The research unit employs environmental specialists and is supported by an Advisory Committee of environmental experts. The research is guided by a criteria paper entitled "The Assessment Process for Green Investment," which sets out parameters for analyzing company policy and management, production processes, and products. In addition to individual company research, the unit conducts surveys of industrial sectors in order to identify best performers. (Surveyed sectors include the electricity distribution companies, the water utility companies, supermarkets, and gas supply companies.) The unit has also achieved numerous successes with companies, such as the publishing of environmental data not previously in the public domain, the cessation of purchases of an environmentally damaging product, and the investment in new manufacturing equipment that does not use chlorofluorocarbons.

As environmental investment continues to gather momentum, the range of support services is growing. There are now brokers offering specialist services such as Alex Brown and First Analysis in the United States and James Capel, Paribas, and Banque Sarasin in Europe. These have been complemented by an increasing number of publications and, perhaps most significantly, the first detailed environmental assessments of corporations, put out by such U.S. organizations as the Council on Economic Priorities and the Investor Responsibility Research Center.

Tessa Tennant, head of Merlin research and director of JTM, believes that for the private investment community to increasingly contribute to sustainable development, further changes in legal and operating practices are needed. These include:

• an improvement in the amount and quality of environmental information about companies available to the financial community (such as improved access to regulatory data bases covering emissions data, health and safety data, and other indicators of corporate environmental performance, and requirements for fuller reporting on environmental management issues in company reports);

• environmental accountability from the fund management community and institutional investors, as defined, for example, in their policy statements, investment criteria, and research initiatives; and

• the introduction of personal savings incentives that encourage investment in environmentally responsible companies (such as for the individual, tax beneficial savings schemes, and for institutions, tax advantages for those willing to hold shares for longer periods).

Lessons Learned

• Environmentally related financial instruments can offer investors competitive investment performance while contributing to sustainable development.

• The equity markets, along with banking, insurance, and development capital specialists, are the four cornerstones of the finance industry from the viewpoint of sustainable development. Each must develop a substantial response to the agenda if the power of the world's capital markets is to be harnessed constructively.

Contact Persons

Tessa Tennant and Mark Campanale
The Merlin Research Unit
Jupiter Tarbutt Merlin Ltd
197 Knightsbridge
London SW7 1RB, UK

15   Managing Cleaner Production

Developing cleaner production processes has been at the heart of the environmental management debate for many years. After two decades of trial and error, there is increasing evidence that business, governments, and the public at large realize that pollution prevention and resource conservation are the most cost-effective ways of achieving a quality environment. The five cases in this chapter illustrate some key elements of this approach. Developing cleaner production processes is only part of the total task of reducing environmental impacts throughout the entire life cycle of a product. The second part, cleaner products, is the subject of the next chapter.

Experience has demonstrated the effectiveness of explicit corporate programs designed to reduce the generation of waste at its source. Case 15.1 describes how Dow Chemical has become one of the leaders in the United States with its Waste Reduction Always Pays initiative, which draws on top management commitment and employee involvement to reach ambitious waste reduction targets.

The main driving force for cleaner production in many developing countries is not government regulation but a resource crunch—in other words, expensive and uncertain access to necessary raw materials for production. Harihar Fibers, described in case 15.2, has achieved considerable cost savings as well as reduced environmental impacts through an aggressive resource productivity program.

Reducing the amount of energy used in production is often the most effective contribution a company can make to sustainable development, particularly in energy-intensive sectors like cement. Case 15.3 looks at how Holderbank drew on its experience with Total Quality Management to introduce an innovative energy reduction scheme that uses the motivation and expertise of its staff.

Multinational corporations have a responsibility and an opportunity to introduce "best available" environmental technology and techniques in their developing-country operations. In Ciba-Geigy's case, described in case 15.4, this has meant introducing an almost "zero pollution" dyestuff production plant in Indonesia, which has paid for itself in the space of two years.

What one person considers waste can be another person's raw material. Case 15.5 considers how ConAgra developed a way of transforming waste from its food processes into a degradable plastic-like material that it will produce in collaboration with Du Pont.

Case 15.1
Dow Chemical: Making Waste Reduction Pay

Many companies still regard waste and emissions reduction as something they must do to satisfy someone else rather than as something they can do to realize economic gain. As a result, end-of-pipe treatment continues to dominate the industrial mindset. This approach can be inefficient both economically and environmentally. Often the most cost-effective way to reduce waste and pollution is to go upstream into the production process and eliminate the problem at its source.

But "much more is at stake," says Dow president and chief executive officer (CEO) Frank Popoff. "Surveys show that the public is skeptical of industry's efforts at environmental protection. If we fail to take the initiative, the result will be a regulatory crunch that costs us—and the public—dearly without achieving significant benefits. Through pollution prevention, industry can be viewed as part of the solution, not as part of the problem." Dow's Waste Reduction Always Pays (WRAP) program clearly shows the mutual benefits for the corporation and the community that can arise from taking the environmental initiative on pollution prevention.

Dow Chemical Company is one the world's leading chemical manufacturers, with 1990 sales of $19.7 billion and operations in 32 countries. The seeds of Dow's WRAP program were planted in the late 1960s under the chairman, Carl Gerstacker. As he said at the time, "The Dow approach to environmental problems is quite simple. Its basis is that the best solution is not to produce the pollutants in the first place." This conviction led to the establishment of a yield improvement program that analyzed the company's manufacturing processes, looking for ways to maximize resource use efficiency: by using fewer raw materials and less energy per unit of product, less waste was generated.

Dow's approach made a qualitative leap forward in 1986 with the introduction of a fully fledged pollution prevention program, WRAP. Mounting public concern following disasters such as the leak of poisonous fumes at Union Carbide's Bhopal plant and the introduction of innovative emissions reporting requirements with Title III of the 1986 Superfund Amendments and Reauthorization Act (SARA) prompted a number of leading U.S. companies to formalize and upgrade their pollution prevention efforts. (See case 11.2.)

As part of its WRAP program, Dow adopted the standard management hierarchy, placing waste reduction at source at the top of the list. Next on the list is the recycling and reuse of waste by-products. When this is not possible, the waste should be treated biologically or chemically, or incinerated. Only when all these options have been tried should the residual waste be landfilled in a secure landfill. One result of the WRAP program is that less than 1 percent of Dow's hazardous waste now ends up in a landfill.

The WRAP program started in Dow's U.S. operations and is now being extended throughout the company's global operations. Each of Dow's five operating divisions in the United States has a designated WRAP coordinator, responsible for implementing the program in a manner most appropriate to the site. All Dow's plants are required to develop an inventory of process losses to the air, land, and water so that they can measure and track their performance. These losses are then calculated as a ratio of production rates, to account for differences in production, to arrive at a weighted average for each facility. The result is an index of pounds of waste versus pounds of product that can be tracked and evaluated by each site on a regular basis.

The program is driven by a combination of senior management commitment and employee initiative. Each year Dow finances the development and implementation of waste reduction projects proposed by its employees. For projects to qualify for funding, they must demonstrate a measurable reduction of waste released to the environment. Projects can involve changes to capital equipment as well as changes to maintenance, operational, or administrative procedures. Ideally, WRAP projects save the company money, but Dow realizes that some projects may not offer a return on investment that can be quantified. At the same time, it is often possible to achieve reductions in pollution without any capital expenditures.

Employees are encouraged to seek out waste reduction projects and are appropriately rewarded for their efforts through personal recognition from peers and management. Each year, outstanding waste reduction projects are selected at each U.S. operating division and the employees involved in their design and implementation are honored at special events attended by their colleagues and senior management. A major goal of the WRAP program is to broaden the thinking of employees so that they look beyond individual plant boundaries and understand the total environmental impact of their chemical processes. They must never assume that some other part of the organization will handle their waste streams.

In five years of operation, the Waste Reduction Always Pays program has lived up to its name. The payoffs have taken many forms. Materials previously lost are recovered and reused, yields are increased, there are savings in transportation and disposal costs, and emissions have been significantly reduced. Joel Hirschhorn, a leading pollution prevention expert, sees WRAP as an example for others to follow: "Dow has had remarkable success at many of its plants, often reducing its generation of all environmental pollutants to levels that few technical specialists thought possible." Specifically, Dow has cut its air emissions in the United States at least in half since 1985.

Team effort from motivated employees has been the key to success. To keep the search for waste reduction opportunities uppermost in the minds of the company's engineers, Dow has armed each with a book of waste reduction strategies, including a list of nuts-and-bolts ideas. Numerous waste reduction teams, involving participation from various sectors, have designed and implemented hundreds of successful projects. "It's always exciting and feels good when the work you do ends up being positive for the environment, and saves the company some money at the same time," says Bill Vanderkooi, a scientist who helped design a successful WRAP project at Dow's Michigan Division acrylamide monomer plant. Jim McKay, a process specialist who worked with Vanderkooi on the project, adds, "It just makes common sense not to throw things away that can be reused. We do it all the time at home with food. Why not do it at work also?"

At Dow's Western Division in Pittsburg, California, a project team representing research, the technology center, the analytical department, and the agricultural chemicals plant identified opportunities to recycle and improve the control of a reactant used in the production of agricultural products. Traditionally, the reactant had been incinerated after a single use. Using computer modelling techniques, the team was able to find ways to cut the consumption of the reactant by 80 percent through recovery and reuse, thereby eliminating 2.5 million pounds of waste per year. As a result, the team saved Dow $8 million annually from reduced raw materials use and lower environmental and labor costs.

One mile from corporate headquarters at the Michigan Division facility in Midland, a waste reduction team at the anion exchange resins plant modified the production system to recover methanol that had previously been sent to the waste treatment plant. Although this cut process losses, it also meant a reduction in plant capacity. So the team identified the additional equipment needed to return to original capacity, while still saving the methanol. Together these changes reduced methanol waste by 660,000 pounds a year, translating into incremental product savings worth $59,500. In addition, the plant is saving approximately $30,000 a year from reduced wastewater treatment costs. "It took a few months to get things where we wanted," says Liane McGill, an engineer who implemented the project. "It took a lot of hard work. But everyone was committed to making it work and it went really well."

On the basis of the success with its WRAP program, Dow has publicly supported regulatory and industrywide efforts to spread the pollution prevention approach. The company responded positively to U.S. Environmental Protection Agency Administrator William Reilly's challenge to industry to voluntarily reduce emissions of 17 high-priority chemicals from 1988 levels by a third by 1992 and by 50 percent by 1995 (known as the 33/50 Program). As well as cutting emissions of these chemicals, Dow has committed itself to cut in half by 1995 emissions of all chemicals reported under SARA Title III, amounting to 121 different substances. Reilly welcomed Dow's positive approach, saying "it's encouraging to see an industrial leader that is ready to step up to the plate as a good citizen to help reduce troublesome pollutants." In addition, Dow is participating in the chemical industry's Responsible Care program, which requires companies to follow a number of guiding principles and codes of best practice, including one on pollution prevention. Under this code, companies have to establish a substantial, long-term downward trend in waste generation and releases to the environment. (See case 12.5.)

Dow has also led the way in developing innovative ways of communicating its efforts to the community. Industry has a responsibility to provide the facts about its environmental performance to the full range of stakeholders who have an interest in its activities. This includes employees, local communities, investors, government, and the environmental community. As well as holding open houses and plant tours, Dow has established community advisory panels (CAPs) at several of its U.S. and other locations. These panels are composed of representatives from the community and meet regularly with Dow personnel to discuss issues such as emergency preparedness, hiring policies, and environmental performance.

Ben Woodhouse, Dow's corporate director of global issues, admits that CAPs were initially seen by some plant managers as unnecessary interference. "Now," he says "they are proving to be a definite benefit to our industry and certainly to Dow. They are bringing the community and the company together through dialogue—not confrontation." Dow recently took the CAP concept one step further by forming a Corporate Environmental Advisory Council, an external group of global policy and opinion leaders, which meets three to four times a year to advise the company on environmental, health, and safety issues. "Through the Council, we hope to broaden our perspective, become more responsive to public opinion, and elevate our performance," says David Buzzelli, Dow's vice-president and corporate director of environment, health, and safety.

Lessons Learned

• Waste reduction should be considered an opportunity, not a burden.

• Employee involvement is essential; successful programs depend on the creativity and enthusiasm of employees.

• Open communication with stakeholders is both a responsibility and the source of numerous benefits for companies.

Contact Person

Joe Lindsly
Manager, Waste Reduction
2030 Dow Center
Midland MI 48674 USA

Case 15.2
Harihar Polyfibers: Promoting Productivity to Prevent Pollution

The need for industrial efficiency is particularly urgent in developing countries, due to increasing pressures on resources, both material and financial—what Prasad Modak, a Bombay-based expert on cleaner production, has called the "resource crunch." Population growth and environmental mismanagement mean that the supply of basic inputs such as water and wood can become uncertain or subject to price rises. In parallel, developing-country governments are progressively reducing subsidies on water and energy consumption while increasing the use of economic instruments to make producers pay the full environmental costs of their consumption. Furthermore, developing countries cannot afford to waste financial resources on costly cleanup programs; pollution prevention is thus a necessity for environmental and economic reasons.

The textile industry is one of the largest in the world in terms of production volumes and employment. The past decade has seen a marked redeployment to the developing world to take advantage of cheap labor and lower operating costs. The production of textiles can be highly water- and chemical-intensive, resulting in considerable emissions of polluted wastewater. In the case of Harihar Polyfibers in southern India, however, resource constraints were turned to both financial and ecological advantage.

Harihar Polyfibers is a medium-sized enterprise, employing more than 1,600 workers, situated on the Tungabhadra River in Karnataka state. It is part of the GRASIM industries group, and produces more than 60,000 metric tons of rayon-grade pulp each year, a staple product of the Indian textile industry. The main source of raw material for Harihar's rayon production is wood pulp, derived from local eucalyptus and casuarina plantations.

In the early 1980s, Harihar began to experience serious difficulties with its pulp supply. The Karnataka state government had decided simultaneously to restrict the quantity of wood from public plantations, because of problems with deforestation, and to increase price by restricting supply. Harihar could not pass on these price increases to its customers in the fiber processing industry because of fierce competition from imported fiber producers. The company's management realized that the only way to survive was to take a hard look at its production processes and radically reduce its costs.

GRASIM's senior executive president, Shailendra Jain, was appointed to lead this productivity drive. Jain is a powerful advocate of the need for improved efficiency in Indian industry to meet the competition of the global marketplace: "It is well known that in our heavy industry sector, the consumption of raw materials, energy, and labour is much higher than in other countries. This not only makes our products more costly...[it also] manifests itself in faster depletion of our resource on one hand and pollution of environment on the other." As a result, Jain believes that "productivity should become as sacred to us as our religious faiths."[1]

Jain implemented his beliefs at Harihar in 1983 with the launch of a four-point plan to reduce production costs through enhanced resource efficiency. The plan focussed on:

• motivating and involving company employees to identify and correct wasteful and inefficient operations through "good housekeeping" practices, and ensuring that capital equipment was properly maintained to increase reliability and resource conservation;

• identifying opportunities for recovering and reusing waste energy and materials;

• targeting the production process itself, and analyzing opportunities for reducing the consumption of chemicals by reducing the number of process steps or introducing innovative technology; and

• installing new equipment with higher intrinsic process/energy efficiency.

To get the program under way, Jain asked all department heads to measure their resource flows and suggest options for improving efficiency. In all, more than 200 projects were implemented between 1983 and 1989, resulting in a considerable reduction in the chemicals and

energy used, as well as a corresponding drop in pollution. Three innovations stand out in particular.

Process Modification. Countercurrent washing is becoming a well-
established method of reducing water consumption in the textile industry. With this process, the least contaminated water from the final wash
is reused for the next-to-last wash and so on until the water reaches the
first wash stage, where it is discharged for treatment. At Harihar, a three-
stage countercurrent process had been in operation for many years. But
the system leaked and required considerable amounts of chlorine to deal
with impurities, which added to the pollution from the process. The
solution that Harihar chose was to add a fourth washing stage. Not only
did the company save nearly 20 million rupees ($770,000) from reduced
chemical inputs, it was also able to cut its energy bill by another million
rupees. In addition, the chemical oxygen demand of the wastewater fell
by 40 percent.

Resource Recovery. At another process at Harihar, wastes produced
during cleaning were sent straight to the sewer, causing blockages and
the loss of valuable fiber. To cope with this problem, a vibrating screen
was introduced to drain the water from the sludge, which was then
bagged and sold to board and cellulose powder manufacturers. As well
as solving the waste disposal problem, more than 100,000 rupees ($3,850)
were made each year from the waste sales.

Energy Conservation. At the end of the production process, the pulp
is drained to remove excess water, flash-dried, and then baled. To cut
down on the amount of energy used, two additional presses were added
to squeeze more water out before the drying stage. This meant that the
drying temperature could be cut from 300 to 200 degrees Celsius. Fuel
oil consumption was reduced, saving 3.5 million rupees ($135,000) a
year. In addition, product quality was improved, since the reduced
drying temperature meant that there was no longer any risk of "burn-
outs."

The overall impact of the productivity program is impressive. Although production increased by almost 20 percent between 1983 and
1989, energy consumption fell by 60 percent, chemical consumption
dropped by 55 percent, and the effluent load was cut by 55 percent. The
quality of the Tungabhadra River downstream of the plant was improved, as was ambient air quality. Although the total program cost
almost 1.8 billion rupees ($69.5 million), the payback period was under

two years at "then costs." Payback on the good housekeeping investments was achieved in just over a year.

Harihar has become a cleaner, leaner company and reduced its dependence on its raw material supply. As Jain told *Business India* in 1990, "we've saved so much by adopting pollution control measures that it's been a blessing in disguise."[2] Importantly for a company in a developing country, these improvements were achieved using local skills and technologies; no foreign consultant was hired and no equipment was imported.

Harihar has also been able to use its experience to develop a consultancy business, improving the productivity of similar companies in India. Five of the Harihar initiatives have been logged in the case study section of the U.N. Environment Programme's International Cleaner Production Information Clearing House, a computer-based system accessible by modem from anywhere in the world that now contains more than 500 examples of industrial best practice. The Harihar experience is thus readily available to stimulate and encourage other entrepreneurs to improve their environmental performance.

During the program, the most difficult issue to resolve was gaining the dedication of the work force. By demonstrating top management commitment to the program, Jain was able to win the confidence of his employees. This led to more than 150 good housekeeping improvements, which in turn formed the stepping stone for other changes.

Despite its improvements, Harihar has not been able to avoid criticism for its environmental performance. During 1988, a controversy erupted over the company's emissions to the nearby Tungabhadra River, following a report alleging "serious pollution." Although Harihar was meeting government regulations, its emissions of an unpleasant odor and strongly colored wastewaters led to a community backlash and a court case. Harihar has now installed a comprehensive wastewater treatment plant. However, it still has not established a separate division dedicated to resource conservation or environmental management issues.

This incident shows the limitations of relying solely on a productivity-driven program. As well as improving resource efficiency and meeting government regulations, companies in a sustainable world must ensure that they meet and exceed community expectations of good environmental stewardship.

Lessons Learned

• Companies can achieve important environmental improvements by introducing comprehensive resource efficiency programs.

• Resource efficiency is only one element of a broader commitment to sustainable industrial development; accountability to community expectations is another.

Contact Persons

Shailendra Jain   Prasad Modak
Senior Executive President   Centre for Environmental
GRASIM   Science and Engineering
Kumarapatnam   Indian Institute of
81 123 Dharwad, India   Technology, Powai
                                        400 076 Bombay, India

Case 15.3
Holderbank: Making Cement with Less Energy

The production of a number of key industrial materials, such as aluminum, steel, and cement, consumes considerable amounts of energy. Price rises following the two oil shocks in the 1970s accentuated existing competitive pressures to minimize energy usage wherever possible. The results in a number of sectors have been promising. In Sweden, for example, the steel industry reduced energy demand in their already efficient processes by 27 percent between 1976 and 1983.

Companies in these energy-intensive sectors now face a range of additional pressures to be more energy-efficient. Numerous studies have identified large energy savings through the adoption of cleaner, more efficient process technologies. But these invariably involve new and extra capital expenditures, which are often not available. Another way to improve energy efficiency is to focus on the "software" side, by building up the skills and motivation of managers and workers to make products more efficiently. This is the approach taken by the Swiss-based worldwide cement manufacturer Holderbank.

Each year the Holderbank Group of companies produces more than 40 million metric tons of cement, 42 million metric tons of aggregates, and

11 million cubic meters of ready-mixed concrete in 24 countries around the world. Making cement is a highly energy-intensive business, and electrical and thermal energy costs amount to 40–50 percent of total production costs. Rising energy prices as well as increasing concern about the environmental impact of its operations led Holderbank deputy chairman M.D. Amstutz to launch an innovative energy conservation program in 1990: "Ecological concerns are a priority in our industry and the reduction of the energy content in cement is a prime target in Holderbank's policy," says Amstutz.

Holderbank engineers have identified five main factors that determine a cement factory's energy productivity: market conditions (such as market prices, rules, and regulations), technology, materials, organization, and people. In the past, the main focus was put on improving process technology and the consumption of raw materials. As a result, Holderbank believes that in most cases its technology and materials are state of the art. The greatest potential for improvement now lies in changes to the external conditions, company organizational methods, and people. This means achieving a new state of the art on the software side of the energy equation by introducing appropriate management techniques and deploying trained and motivated personnel to ensure the company taps the full potential of its technologies and materials. As one Holderbank manager commented, "Our goal is to ensure that the people are both sufficiently aware and capable to find and implement the best relationship between their factory and the environment."

Building on its successful experience with Total Quality Management techniques, Holderbank established a Better Cost Management program in 1990 to focus attention on the need to reduce production costs; energy conservation was seen as the top priority for action. The first step was to draw up groupwide targets for energy use per unit of product for the rest of the decade. Next, a six-stage program was designed, to be implemented in phases throughout the decentralized Holderbank group. A small four-person team from the group's consultancy company, Holderbank Management & Consulting Ltd (HMC), would provide technical advice and assistance that is paid for by participating companies. At each designated company, the existing situation and the potential for savings would be analyzed by local and HMC staff. Conservation targets, expressed in terms of kilowatt-hours per metric ton of cement, would then be defined, laying the basis for a comprehensive action plan. Project planning would be aided by the introduction of a computer-aided energy management information system.

At each plant chosen, the following six stages would be followed:

Stage 1:   Preparing and undertaking a survey at the plant by local management and the HMC.

Stage 2:   Preparing an energy conservation workshop.

Stage 3:   Conducting the workshop.

Stage 4:   Introducing energy control devices and organizational methods.

Stage 5:   Training managers and personnel. During this period the plan of action is put into practice. The measures in the plant are accompanied and supervised by HMC.

Stage 6:   Holding a follow-up workshop at the plant.

To test this methodology, Holderbank chose two pilot plants at its Mexico operations. Within five months the first pilot plant had saved electricity worth $280,000 and the second, $160,000, due to enhanced commitment, awareness, knowledge, and management methods. The company also learned a number of lessons that have been fed back into the program.

In particular, the HMC group overseeing the pilot plants realized that more attention was needed to stimulating employee involvement in the program. To do this, Holderbank came up with the idea of "energy circles." Based on the quality circle idea, these energy circles would be established to help motivate and train staff to seek out energy conservation opportunities, especially in the shop-floor manufacturing units. The establishment of energy circles has been added to Stage 3 of the program. The pilot projects also led to an expansion of Stage 5 to include a more systematic review of progress at the plant by HMC personnel. The plant is now required to file monthly energy reports to the HMC, which provides comments and advice for further progress. In addition, HMC staff make bimonthly visits for the six-to-nine-month period after the start of the project to monitor implementation.

According to Holderbank, the advantage of the staged introduction is that measurable savings can be found quickly, leading to employee motivation and allowing the company to make modifications where necessary. Holderbank also stresses the importance of combining local commitment, driven by a mission statement written by the plant management, with a permanent energy management organization at both the plant and corporate level. Each plant now has an interdisciplinary energy management team drawn from all the relevant departments and an energy management coordinator, reporting to the local chief executive. The program is being implemented in five other plants, with plans to increase this to 15 plants in Latin/Central America and Spain before the end of 1992. A kick-off meeting was held in May 1991 in Mexico City, lead by Holderbank chairman Thomas Schmidheiny and deputy chairman Amstutz and attended by CEOs and staff from the 15 target companies.

Although many of the solutions for improved energy productivity that emerge during the program often seem trivial, collectively they can have a great impact. According to Holderbank executives, this is because people are working in an interdisciplinary fashion, using techniques such as group dynamics and brainstorming to generate new ideas and solutions. Claude Brunner, an energy coordinator at one of the pilot plants before joining the energy management team at HMC, has been pleasantly surprised by the success of these techniques: "It's amazing how much employee involvement and the system of continuous improvement can contribute to reducing the energy content of our products." Based on initial results from the pilot plants, Holderbank believes it can achieve energy savings of 3–13 percent within a year with little or no investment; total savings of 17–28 percent will require more time and money. The program has been so successful that companies are now coming forward to volunteer themselves.

Lessons Learned

• The systematic introduction of an energy management program, involving all employees, can generate significant savings with low capital investment. This is a key message for small and medium-sized companies in both industrial and developing countries.

• Energy savings generate profits and reduce adverse environmental impacts.

Contact Person

Michael Blanck
Holderbank Management & Consulting Ltd.
CH-5113 Holderbank, Switzerland

Case 15.4
Ciba-Geigy: Designing a Low-Pollution Dyestuff Plant

Many leading multinationals are already designing and operating production facilities that meet the same high standards throughout the world. At the UK chemical concern ICI, for example, the corporate commitment to minimize environmental impacts has been translated into the design of a new terephthalic acid plant in Taiwan that will "virtually eliminate the production of waste," according to the group director responsible for the environment, Christopher Hampson.

Ciba-Geigy is a Swiss-based multinational manufacturing firm that produces pharmaceuticals, agrochemicals, dyestuffs and textile chemicals, plastics, pigments, and vision care products. In 1990, consolidated sales from the group amounted to SFr19.7 billion ($14.1 billion). The nature of its business and its commitment to high environmental standards has meant that approximately 17 percent of total capital expenditure is now devoted to environmental investments. Ciba-Geigy operates in more than 80 of the world's 120–140 developing countries, and the firm has pledged that "in environmental protection, the objectives we pursue in the Third World are the same as those we pursue in the industrialized countries."

In practice, the means by which the firm's Third World subsidiaries meet these objectives will vary. For example, Ciba-Geigy requires that all its facilities have drain covers to capture accidental spills or leaks. However, the costs of specially designed covers—basically partially filled bags—were prohibitive for the firm's Atoto plant in Mexico. As a result, the environmental officer there applied some lateral thinking and now uses large inner tubes instead. Not only are the inner tubes very inexpensive, they also perform better than the specially designed covers.

The solution to a potential wastewater pollution problem at the Candra Sari dyestuffs plant in Jakarta, Indonesia, was similarly innovative. Ciba-Geigy had established a facility to manufacture textile chemicals in the mid-1970s. In 1985, it decided to add a standardization plant, which would process imported crude dye cake and produce dyestuffs for the growing Southeast Asian textiles market. But the very quality that made the final product so desirable—its high color intensity and stability against a range of chemical, physical, and biological effects—would have made untreated effluent from the production process impervious to normal biological treatment. The company recognized that simply to discharge the liquid into the Candra River would be unacceptable. However, to install a treatment system that could remove the color from the effluent would cost more than the standardization plant itself. As W. Schaad, technical coordinator for regional production management, says "a low-cost method to reduce the effluent loads and color had to be found."

Engineers from Ciba-Geigy headquarters in Basel worked with local staff to design the new production facility so that no untreated wastewater was discharged to the river. The team determined that the cleaning process required between each new dyestuff was the source of the highly colored water. Since the unit produces a variety of different color dyes, the plant has to change over on average twice a week. This involves a complete clean out of the production unit to prevent any cross-contamination of the dyes.

Instead of discharging this water, the firm decided to collect and store it, and to recycle it the next time the dyestuff is being produced. To minimize the storage of this wastewater, the company also designed the process to minimize water use. Ciba-Geigy has found that it has had to keep about 300 metric tons of water in storage, which takes up one third of the plant space. Yet this "closed loop" production system has enabled the company to recover 3 percent more dye from the water, which within only two years covered the costs of the necessary investments. As a result, says Schaad, "the company made a more useful product with minimal pollution."

Robert Unseld, managing director of the Candra Sari complex, stresses that very tight control and good management of the water collection and storage operations are required. He believes that "the key factor in achieving success with this system has been the continuous and intensive education of management and staff in order to increase their environmental awareness." The company has since applied similar water recycling procedures to the original textile chemical production plant, resulting in a reduction in chemical oxygen demand in its wastewater effluent of more than 80 percent between 1988 and 1990, despite a 25-

percent increase in production volume during the same period. Furthermore, the actual volume of wastewater discharged from the plant has
been reduced by at least a third, from more than 0.3 cubic meters per
metric ton of product to less that 0.2 cubic meters.

The Candra Sari plant now operates according to environmental
standards that would meet the most demanding European requirements. With minimal investment, the company has eliminated a potentially serious pollution problem. Ciba-Geigy executives believe that
similar achievements could be made by "many small production facilities located throughout the developing world."

Lessons Learned

• Multinational corporations can benefit from pursuing the same environmental objectives at all their plants regardless of local regulatory
requirements.

• There are low-cost solutions that can in fact act as a stimulus to
designing more efficient and less polluting plants.

Contact Person

Janet L. Ramp
Environmental Protection
Ciba-Geigy
Klybeckstrasse 141
CH-4002 Basel, Switzerland

Case 15.5
ConAgra/Du Pont: Profiting from Recycled Waste

"Imagine wastes being converted into high-value packaging, which
offers a viable substitute for several petroleum plastics, subsequently
degrades totally, or, alternately, can be recycled back to the exact same
use," says Mark Montgomery, president of Ecochem, a new joint venture
established to do just that. Ecochem was formed in late 1990 by ConAgra,
Inc., the world's fourth largest food company, and Du Pont, the world's
largest chemical company. It is a unique venture in that it takes a waste
product from ConAgra and applies the materials knowledge of Du Pont
to develop a biodegradable plastic-like material, thus tackling two

environmental problems—production and packaging wastes—at once.

Ecochem will commercialize innovative technology to produce high-quality lactic acid and polylactides from cheese waste or corn by-products. Lactic acid is primarily a food additive and polylactides are an innovative material-like plastic that easily degrades into water and carbon dioxide. Although the environmental benefits of the technology and the strength of the parent companies suggested that a potentially huge commercial success could be achieved, the birth of the new company was a painful one.

ConAgra United Agri Products Company and Edward Lightfoot of the University of Wisconsin developed the basic technology to produce lactic acid from cheese whey permeate, a waste product from cheese production. Lactic acid can be used to produce polylactides, which, unlike petrochemical-based plastics, are easily degradable in the presence of moisture, air, and common bacteria—and thus an excellent material for degradable packaging. ConAgra realized that it lacked the engineering and manufacturing resources to start a business to exploit the breakthrough. Since Du Pont was pursuing polylactide technology for medical and packaging uses and since ConAgra and Du Pont were already involved in other ventures together, teaming up to develop this technology was an obvious way forward. Du Pont's polymer engineering and manufacturing expertise, combined with ConAgra's access to raw materials and markets, promised to be a strong combination.

Ecochem will use these polylactides to develop packaging for a range of sectors, including grocery products, fast-food packages, personal hygiene products, and insulated drinking cups. For Montgomery, "green packaging" is much more than a marketing ploy. "As people around the world pay more attention to improving the environment," he says, "companies should uncover and nourish opportunities to develop new businesses that respond to this need, and add to the bottom line." In addition to their degradability, polylactides could also be applied to facilitate greater paper recycling rates by allowing the development of printing products and coatings that can be removed easily with existing paper recycling technologies.

As a result of the degradability and recycling benefits, Montgomery believes that polylactides could eventually reduce significantly the solid waste going into landfills. "This environmental benefit, coupled with the fact that polylactides are produced from renewable resources, makes polylactides a huge environmental success," he says.

Ecochem still has a long way to go, however. A pilot plant producing the polylactides started operation in 1992 to prove the nonpolluting manufacturing processes and serve as the basis for commercial polylactide plant design. The output of this pilot plant will also be used in product prototypes to refine market size estimates and confirm economic justification for large-scale investment. Construction of a $20-million lactic acid facility has begun. A world-scale integrated polymer plant should be completed and operational by late 1994. With a total investment of more than $100 million, Ecochem is expecting that sales of degradable packaging could reach several hundred million dollars in the second half of the 1990s.

Building the Ecochem joint venture required considerable personal efforts to accommodate differing internal management practices of both parent companies. Du Pont was spending large sums of money to support polymer recycling, and some individuals feared that degradable polymer research funding could detract from the success of the recycling effort. The characteristics of polylactides that enable recycling and their subsequent compatibility with overall solid waste reduction was not immediately apparent. But other Du Pont personnel stuck their necks out to champion polylactides and were able to overcome this resistance. At first, research had to be funded out of general research funds, but when test results proved positive and the inexpensive source of lactic acid from cheese whey was discovered, stronger research and development funding became available. Seeking to speed lactic acid and polymer process development, Du Pont personnel were directed to move the technology inexpensively to a point where potential economic viability could be reasonably assessed.

The staff involved in the Ecochem venture came from very different companies. The initial Du Pont "champions" were essentially engineers, eager to exploit the new degradable polylactide technologies, while the "champions" within ConAgra were entrepreneurs with a strong bias toward low-cost production of commodity products. The joint venture also had to accommodate some differing approaches of the parent companies. Traditional Du Pont wisdom about new technology development and commercialization is that "good science and engineering" requires large research and development spending over a long time. ConAgra, however, has an investment creed of carefully controlling capital and achieving their stringent 20-percent return on equity standard in a relatively short period. Thus the partners had to reconcile very different perspectives on investment levels and payback schedules. In the early stages of getting the joint venture under way, existing managers in each company handled the project in addition to their ongoing responsibilities. Although this is not unusual, development of polylactides would have been advanced by several months if sole responsibility had been given earlier to a key manager.

Executives at both Du Pont and ConAgra have learned a number of lessons from the Ecochem experience. Special effort is needed to turn sustainable development opportunities into profitable reality. New forms of business alliances may be required. Furthermore, while the two companies believe that sustainable business projects can be profitable, returns may not emerge immediately. As a result, sustainable development projects should be given the benefit of the doubt in borderline cases.

Ecochem has shown that the potential impact of converting renewable raw materials to high-value products is too great to ignore or treat "in the normal course of business." Companies should be alert to projects with significant profit and sustainable development potential and consider some mechanism that would elevate them within company management for special consideration for funding and human resource support.

Lessons Learned

• The use of renewable raw material feedstocks, degradable materials, and the development of "waste elimination" manufacturing processes are increasingly realistic for profitable business development.

• The commercial opportunities that eco-efficiency opens up may require resources beyond the capacities of a single company; combining forces in joint ventures may offer the best potential in these instances.

• New decision-making mechanisms may be needed to give priority and support to projects supporting sustainable development.

Contact Person

Mark Montgomery
President, Ecochem
3411 Silverside Road
Wilmington DE 19898 USA

16   Managing Cleaner Products

The late 1980s saw an unprecedented surge of environmentally related product development. Governments and consumers were both demanding products that could be produced, used, and disposed of with minimal environmental impacts. These pressures have sent many companies back to the drawing boards to design products that inherently pollute less and consume fewer resources. This chapter contains eight cases that illustrate some of the key aspects of cleaner product development.

The increasing complexity and international nature of product sourcing means that companies that want to act in a responsible fashion have to adopt a global approach to product stewardship. Smith & Hawken undertook an extensive research program, described in case 16.1, to assess whether the tropical hardwoods it used originated from forests managed sustainably. When it found that they did not, it introduced a program that not only improved its own products, but led to the establishment of an industrywide screening process for tropical timber.

Trying to assess a product's environmental impacts from its conception through to its disposal is a complex process, involving many different actors in a number of countries. Over the years, life-cycle analysis has been developed to address this problem. Case 16.2 looks at how Procter & Gamble has become a recognized leader in the discipline, applying it to a range of its products and packaging, including that for detergent conditioners. This helped the company make design improvements that cut the amount of packaging waste its products contributed. Case 16.3 considers how Migros, a Swiss retailer, has used life-cycle analysis in its retail operations.

Increasing scientific knowledge has meant that materials once thought perfect for a particular task have to be removed and replaced with new substances. HENKEL faced this problem with the use of phosphates in its detergents, which became linked with water pollution problems. As case 16.4 describes, an extensive research program enabled the company to introduce a range of phosphate-free detergents and increase market share.

Traditionally, products have not been designed with energy efficiency in mind. Case 16.5 considers how design improvements can yield important savings in a product's energy use, and make it more competitive. To meet a gap in the market, Laing Homes introduced well-insulated timber frame housing in the United Kingdom, saving itself and its customers money while reducing resource use and pollution.

At the end of a product's useful life, it has to be disposed of in an environmentally acceptable way. Declining landfill capacity and concern about the waste of resources prompted Volkswagen and other automakers to assess how they can recycle their cars. Case 16.6 describes a Volkswagen pilot plant that has provided useful guidance for product designers and for an eventual network of licensed vehicle disassemblers.

Retailers perform a crucial intermediary role between producers and consumers. Although attention has thus far been focussed on the schemes introduced by retailers in North America, Europe, and Japan to provide their customers with improved products, companies in developing countries are also taking a stand. Case 16.7 looks at the comprehensive program launched by South Africa's leading chain store, Pick'n Pay, following an environmental audit.

Case 16.8 describes how ENI, a major Italian energy company, developed a substitute product for lead used in gasoline as an octane enhancer. The new product has significant environmental benefits and created new business opportunities for the company.

Case 16.1
Smith & Hawken: Promoting Products of Sustainable Forestry

Based in Mill Valley, California, Smith & Hawken is a mail-order and direct retail company, with 230 regular employees and revenues of $50 million per year. Following initial successes in selling high-quality gardening equipment, Smith & Hawken expanded to outdoor garden furniture in 1984 and to clothing and housewares in 1990.

Its cofounder, Paul Hawken, has made environmental responsibility a company priority; he considers it a matter of personal values as well as good business practice. He currently sits on the boards of Conservation International and the National Audubon Society. Smith & Hawken's environmental activities include partnerships with environmental groups and community outreach, as well as internal, office-based efforts, such as conducting an externally reviewed environmental audit and installing comprehensive recycling programs. The company also has a history of environmentally innovative product development. For example, it has used 650,000 board feet of redwood taken from 100-year-old disassembled wine vats as raw material for furniture and trellises.

The company's sensitivity to environmental issues meant that Hawken was immediately receptive when he began to receive letters in 1988 from customers concerned about the origins of the teak-based furniture in his catalogue. Most teak comes from Southeast Asian countries, where deforestation was reported to be proceeding at rapid rates. Smith & Hawken's customers wanted assurance that the company's products were not contributing to the problem. The teak used by its Thai manufacturer came from Thailand and Myanmar (formerly Burma), which both claimed to be following sustainable forestry practices. But there were rumors of irregularities. Hawken went to Thailand and Myanmar to learn more about harvesting practices, but remained dissatisfied. Rather than simply dropping teak—which could have had a disastrous impact on employment in the Thai furniture business—or doing nothing, Hawken decided to commission a thorough analysis of the situation, with the aim of guaranteeing the sustainability of the forestry practices of its suppliers.

From January to November 1989, Ted Tuescher, who later became Smith & Hawken's environmental director, researched technical, economic, and historic information on forestry in Southeast Asia. He visited the region's wood suppliers and furniture manufacturers to learn where the wood came from, how it was harvested, and how the local communities benefitted from the wood revenues. His report on Thailand concluded that "there is no widespread commitment outside the Forest Department to manage the forests in an environmentally sustainable manner. More effective protection of the forests from illicit removals and conversion to agriculture would be required." In Myanmar, he found the situation was equally unsatisfactory. Moreover, he confirmed rumors that the repressive military regime was increasing sales of logging concessions in order to generate foreign currency to finance its war against insurgents. Although unintentional, Smith & Hawken was indirectly supporting this system.

However, Tuescher's research was not all negative. He visited social forestry projects and furniture manufacturers based in Java, Indonesia. Wood for these facilities came from Javanese teak plantations first established in the 1870s. Tuescher was impressed with the harvesting methods used at these plantations. Representatives of SKEPHI, an Indonesian forest conservation association, supported the harvesting methods used on the plantations and saw long-term benefits for local communities if the government allowed greater community participation in local land management.

Tuescher's 200-page report concluded that Smith & Hawken's existing supplies of teak were unsustainable. The company was faced with a major business and environmental dilemma. Teak furniture accounted for 20 percent of Smith & Hawken's revenues at the time, and changing sources of supply would be expensive and risky. Nevertheless, based on Tuescher's findings, the company required its Thai manufacturer to seek new sources from Java; it also began to evaluate alternative producers in the event that the Thai company failed to change.

While these efforts were going on within Smith & Hawken, deforestation and tropical timber use rapidly became a leading public issue in the United States. A number of conservation groups, such as the San Francisco-based Rainforest Action Network, began to call for a boycott of tropical timber products, including those from plantations. This spurred a fierce debate between the environmental and business communities. However, a conference hosted by the New York-based Rainforest Alliance, which brought together representatives from both sides, concluded that "a boycott, except in specific circumstances, would not promote sustainable forestry." This matched Smith & Hawken's belief that a complete ban would be counterproductive to forest conservation and harmful to Asian and Latin American communities that relied on forest revenues. As Tuescher says, "Instead of leaving the teak wood business, we decided to be a responsible participant. Otherwise a less responsible participant would step in to fill our place as a buyer of teak."

Despite this open approach, Smith & Hawken was criticized in the media for putting business interests before environmental responsibility. *Outside* magazine, for example, claimed that Hawken was dragging his feet on the issue of timber from Myanmar. In the face of a possible consumer backlash against all tropical timber products, Hawken and Tuescher went on the offensive with two initiatives. First, they launched a campaign to educate the public and conservation community on the risks and opportunity of harvesting tropical woods. Tuescher met with conservation groups and shared all the data he had gathered. The company's efforts were rewarded in April 1991 when it received the Environmental Stewardship Award from the Council on Economic Priorities.

The second step was to promote the establishment of an independent certification program for sustainably harvested tropical wood products. Smith & Hawken donated $45,000 to the Rainforest Alliance, which was used to set up the program. Teams from the Alliance are now able to visit facilities around the world to certify that products sold in the United States are manufactured in a sustainable fashion. Ivan Ussach, director of the Alliance's tropical timber project, credits Smith & Hawken with being at the forefront of investigating and verifying sources of tropical timber for U.S. consumption.

Unfortunately for Smith & Hawken, it proved impossible to continue with its Thai manufacturer. Although the manufacturer had agreed to switch to Javanese teak, Smith & Hawken, along with the Rainforest Alliance, soon realized how difficult it was to verify compliance. The Thai supplier was also continuing to use wood from Myanmar for other customers. Hawken decided to transfer production to an Indonesian manufacturer, where it was easier to track shipments of wood.

The first deliveries from the new supplier began to arrive in August 1990, and by mid-1991 all Smith & Hawken's teak furniture was being made in Java. Considerable costs were associated with the change of manufacturers. The teak itself was 5 percent more expensive, while a number of product lines had to be dropped because the Javanese producer did not have adequate equipment to produce curved products. However, the elimination of teak from Myanmar meant that Smith & Hawken is on the way to becoming, in Ussach's words, "a Smart Wood Company," whereby all its wood products are certified to come from reputable suppliers.

To meet the Rainforest Alliance's criteria for this, a supplier will have to minimize environmental impacts, provide social benefits to local communities, and operate according to a long-term management plan. Although thousands of companies in the United States use wood products, only a handful qualified as "Smart Wood Companies" in 1991. Not only has Smith & Hawken stopped using teak from Myanmar, it has also discontinued one of its more popular items, an Adirondack chair made from Honduran cedar, until its manufacturers can secure a source of timber certified by the Alliance.

In all, the research and changes in teak supply cost Smith & Hawken well over $100,000, not including lost business amounting to several hundred thousand dollars. But in doing so, the company has been able to avoid a consumer backlash. Tuescher argues that Smith & Hawken decided "to opt for constructive engagement. If there had been no avenues for constructive engagement we would have left the business." Although it is difficult to judge whether Smith & Hawken has gained market share, it has certainly helped change the rules of the furniture business in the United States, prompting other companies to reexamine their sources of supply. It is this ability to shape the marketplace through an interactive approach to environmental problem-solving that could be one of the key characteristics of companies seeking to gain a new type of competitive advantage consistent with sustainable development.

Lessons Learned

• Companies should include the sourcing of their components and raw materials in their efforts to become eco-efficient.

• Partnerships with environmental groups and local communities are essential to ensure credibility.

• Leading companies can help shape the marketplace through environmental innovations, thus making their competitors adapt to their advantage.

Contact Person

Ted Tuescher
Environmental Director
Smith & Hawken
25 Corte Madera Ave.
Mill Valley CA 92941 USA

Case 16.2
Procter & Gamble: Using Life-Cycle Analysis to Cut
Solid Waste

A new environmental management tool, life-cycle analysis (LCA), offers a way of accounting and comparing environmental emissions and re-source requirements for different product options. Although LCAs have been in existence for more than 20 years, it was only in the late 1980s that they gained widespread recognition, reflecting the increased awareness of the global impacts of individual consumer choices. Business, govern-ments, academia, and citizen groups are now working to improve the techniques and application of LCAs to product stewardship issues.

A 1990 workshop hosted by the U.S. Society of Environmental Toxicol-ogy and Chemistry identified three distinct elements of an LCA. First, there is an inventory of environmental emissions and resource consump-tion at each stage of the product's life cycle, including raw material procurement and processing, manufacturing, distribution, consumer use, disposal, and possible reuse. The second stage involves measuring and assessing the environmental impacts of the emissions and the resource use. The final stage assesses the opportunities for improvement. To date, most LCAs have concentrated on the inventory stage.

A heated debate has arisen about the relative merits of LCAs. On the one hand, enthusiasts have seen them as a panacea for environmental management. On the other hand, critics have chastised LCAs as incom-plete and premature. Despite this apparent polarization, a number of companies are pioneering the use of LCAs as pragmatic management tools to improve environmental performance. One of the leaders in the field is Procter & Gamble (P&G).

As one of the world's largest consumer products companies, selling more than 160 advertised brands in 140 countries, Procter & Gamble is deeply concerned about the issue of how to act as a global steward of its product portfolio. The company has observed a significant increase in consumer interest in the environmental impacts of its products in recent years, and has seen environmental considerations play an increasingly important role in purchasing decisions. To address these concerns and fulfill its wider responsibility to the community, P&G has drawn up a comprehensive environmental quality policy. This recognizes the company's responsibility throughout the life cycle of the product, in-cluding a commitment "to reduce or prevent the environmental impact of our products and packaging in their design, manufacture, distribution, use and disposal whenever possible." The policy also states that P&G will "continually assess our environmental technology and programs, and monitor progress toward environmental goals."

At P&G, risk assessment remains the foundation of ensuring environmental quality. Complying with local and federal regulatory requirements and maintaining effective pollution prevention programs at its production sites are also essential. However, P&G has also pioneered the use of LCAs. Although additional development is needed to realize the full potential of LCAs, P&G believes they can already be effectively applied as a valuable internal environmental management tool. In the past three years, P&G has applied LCA to diapers, to a broad range of detergent and personal care product packaging, and to the sourcing and production of surfactants used in various detergents and cleaning products.

LCAs have provided valuable information and insights about the company's products and processes, as well as those of its major suppliers. This information is then used to help improve designs of existing products and processes, and in the development of new ones. Based on its experience to date, P&G sees LCA as a useful environmental accounting device that can play a unique but by no means all-inclusive role in environmental management. As Charles Pittinger of P&G's environmental safety department says, "No single procedure or approach can alone satisfy all needs; an integrated approach is essential."

Solid waste disposal is a serious issue the world over. The costs of managing the problem are escalating as landfill capacity declines and new facilities become more expensive. Procter & Gamble believes that a variety of things need to be done and can be done to alleviate this problem. According to Pittinger, "LCA can help point the direction for change and serve as a ruler for progress." P&G stresses that it is vital that suggested improvements offer equal or improved performance and value to the customer. In a free market, if a product or package fails in the marketplace for any reason, the environmental benefits it may have offered cannot be realized.

Procter & Gamble has in recent years introduced a number of initiatives worldwide to reduce packaging waste for numerous products, including fabric conditioners, laundry detergents, and hard surface cleaners. In 1989, P&G commissioned leading LCA consultants Franklin

Associates to develop LCAs for a range of different packaging systems used for a variety of cleaning products. All stages of the packages' life cycles were examined, including the implications for secondary packaging (such as boxes to ship cases of the product). Because hundreds of different scenarios were examined, the study took over a year to complete.

Seven different packaging systems were studied for a single P&G fabric conditioner marketed as Downy in the United States and as Lenor in Europe. The options were representative of major types of possible packaging and/or product delivery systems currently used by manufacturers. They included the optimization of raw materials usage, through, for example, increased use of recycled materials, as well as product concentration and the introduction of reusable and/or refillable products. Some of the packages examined are or have been used commercially; others, such as a compostable paperboard carton for refills, require further development before their ultimate feasibility can be assessed.

To ensure comparability, the LCAs of the different options were based on the packaging required to deliver 1,000 liters of single-strength conditioner to the consumer, and included the raw material procurement, fabrication, distribution, and disposal stages. The results were then compared with a standard, nonrecycled 64-ounce virgin high-density polyethylene (HDPE) bottle. P&G was careful to distinguish between energy used to power industrial plants and vehicles, and that used as a feedstock for plastics. It was assumed that the reference container was not recycled or reused following consumer usage. The study made a number of interesting discoveries.

First, the use of soft pouches or paper cartons can make a major contribution to reducing packaging waste at source, resulting in up to a 95-percent reduction compared with the standard plastic bottle. The development of triple-strength fabric conditioner makes such an option possible. Concentration and/or volume reduction is possible for many other product categories, including granular detergents and certain paper products. The use of a concentrated product does, however, require some commitment from the consumer, who needs to transfer the product between containers, and in some instances dilute the product before use. Consumers in Canada, Germany, and the United States appear to be more receptive to this small inconvenience than do those in other countries.

Second, the recycling and reuse of the polyethylene bottles offers minimal benefits. In the case of including 25-percent recycled HDPE in the bottle and encouraging consumer recycling, the amount of wastewater would actually increase, due to the need to clean the plastics before reuse. The recycling option has other drawbacks: it requires consumer commitment, the development of an efficient municipal recycling infrastructure, and the existence of viable end-use markets for recycled materials.

Finally, 25-percent composting of the paperboard could offer additional solid waste benefits even for relatively small paperboard cartons. Solid waste could be reduced 16 percent relative to the noncomposted container. The reduction does not equal a full 25 percent because solid wastes of secondary shipping cartons (such as corrugated boxes) remain the same for both options. For larger volume paper products such as diapers, benefits of composting would be more dramatic. Furthermore, municipal composting can be extended to a broad range of solid wastes including food scraps, office supplies, and yard wastes. It is estimated that in some U.S. communities as much as 60 percent of solid wastes currently being landfilled could be diverted to municipal composting systems, if facilities were developed.

LCA has enabled P&G to develop packaging strategies that match consumer expectations and waste management infrastructures of different markets. In the United States, P&G has been able to incorporate approximately 30-percent recycled HDPE in its Downy bottles over the past two years. More recently, a gabled-top paperboard refill carton containing triple-strength conditioner has been introduced. In Europe, the triple-strength product is marketed as Lenor in a flexible, one-liter polyethylene terephthalate (PET) pouch. P&G is also exploring the use of the paperboard carton in several European countries.

Consumer education and environmental awareness have been vital to the success of the triple-strength conditioner. For example, a triple concentrate version of Downy had been available in the United States years prior to the introduction of the Downy refill, with only marginal consumer response. Increased environmental awareness paved the way for the success of the refill product. A simple statement on the package, "Same Downy Softness and Freshness—Less Packaging to Throw Away," provided performance assurance and highlighted the environmental benefits. In less than a year, the refill had taken nearly 40 percent of

Downy sales in the United States. The response to concentrated products in Europe has been equally strong.

Lessons Learned

• Life-cycle analysis provides an effective internal management tool to identify potential improvements in process and product design.

• Solutions to environmental problems need to be tailored to differing consumer expectations and regulatory conditions. There is no single global optimal product or package.

• Consumer education may be necessary to overcome hesitancy about environmentally improved products.

Contact Person

Charles A. Pittinger
Procter & Gamble
Ivorydale Technical Center
5299 Spring Grove Avenue
Cincinnati OH 45217 USA

Case 16.3
Migros: Using Life-Cycle Analysis in Retail Operations

Migros operates the largest chain of grocery stores and retail stores in Switzerland. In 1991, the company had sales of more than SFr13 billion ($9.3 billion), marketed more than 30,000 different products, and had more than 70,000 employees. Migros also has had an outstanding record of profitability ever since its founding in 1925 by Gottlieb Duttweiler, who imbued the company with his personal philosophy of "social capital" by which he fostered free enterprise concepts with responsibility to promote human well-being and improve the quality of life for individuals.

As a result of Duttweiler's basic social philosophy, it is not surprising that Migros would assume a leadership role in progressive environmental action. The company has adopted a wide range of energy conservation, recycling, and environmental information programs. Although it operates almost entirely within tiny Switzerland, it is well known internationally. The German newspaper *Frankfurter Allgemeine Zeitung* wrote an article entitled, "At the Migros Cooperative You Also Buy A World-View," indicating that the heritage of its founder had earned Migros international recognition for its environmental policies.

Migros has used life-cycle analysis to help its managers make sound business decisions and give environmental factors full consideration. How can you tell accurately which product is most "environmentally friendly?" To do the analysis correctly requires going back to the raw materials input, the production process, the transportation, the packaging of the product, its use, and finally its disposal. Obviously, with 30,000 separate products, this could be a monumental task. The Migros program began with the development of a computer-based environmental information system. Beginning in 1985, this internal information system was used to track individual environmental actions and to generate an annual report to help managers keep track of progress and begin to make comparisons.

That same year Migros began the development of Oekobase 1, a computer-based system to assist managers in the comparison of different packaging materials based on the criteria of energy, air pollution, water pollution, toxicity, and solid waste. The computer could generate graphics to display the relative impacts of different packaging alternatives. It soon became apparent, however, that conducting cross-media analysis was difficult. Product A, for example, might be clearly less polluting of the air but more energy-intensive than product B. As a result, the Migros staff realized that they needed a more "intelligent" system to help resolve these more complex environmental questions.

In 1990, staff began the development of Oekobase 2. This program assigns negative "eco-points" for each adverse environmental impact associated with a packaging system, from resource extraction through final disposal of any nonrecycled waste. In one example, a conventional tin can package for coffee was contrasted with an aluminum-coated paper container: the tin system earned 98 negative eco-points while the aluminum system had only 12. Migros began a consumer education program and now markets all its preground coffee in the aluminum foil packaging, with significantly less adverse environmental impact.

The eco-point system can play an important role in the future as product managers are given targets of not exceeding a certain number of points in their product selection. The overall target is to reduce the yearly amount of eco-points per unit of products sold, starting in 1993.

First attempts are being made to use Oekobase 2 in a noodles factory owned by Migros. A team of Migros specialists and external consultants try to record all use of raw materials, energy, water, and soil, including that used for employees and transport. This input is factored into evaluation of products, waste, and emissions.

Public education and clear consumer information are crucial. The retailers are the gatekeepers between the suppliers and the consumers. But they cannot sell products the consumer rejects. In the coffee packaging case, the consumer accepted the "environmentally friendly" solution, but this is not always true. As Migros president Jules Kyburz says, "We Swiss are very reluctant to changes at first sight; that is also true concerning the environment." Twenty years ago, Migros launched a detergent that was free of phosphates, but nobody bought it. Similarly, 15 years ago the company tried to market unleaded gasoline, but again it was too early to gain public acceptance. "It took 20 years for society to change its mind," Kyburz notes. "But times have changed: environment is now marketing instrument number one." Today the pioneering efforts of Migros are paying off in consumer loyalty, profitability, and steps toward sustainable development. Migros now shares its Oekobase 2 computer program with other retailers at cost.

Lessons Learned

• Life-cycle analysis is a management tool that retailers can use to help their managers make better environmental decisions and improve consumer awareness of environmental issues.

• Retailers can be profitable and environmentally progressive simultaneously.

Contact Person

Walter Staub
Leiter Infostelle Umwelt
Migros-Genossenschaftsbund
Limmatstrasse 152
Postfach 266
CH-8031 Zurich, Switzerland

Case 16.4
HENKEL: Developing Substitutes for Phosphates in Detergents

Getting clothes clean without harming the environment with aggressive chemicals has become a key consumer issue of the 1980s and 1990s. Everyone needs some form of detergent, but increasingly the public does not want cleansing power to come at the expense of damaged waterways. Since the 1950s, the use and effectiveness of detergents has grown enormously in line with the expansion in demand for clothing and increased consumer expectations of detergent performance. Simple soaps and bleaches were replaced by technologically advanced ingredients. Phosphates were one of these new ingredients. They cleaned better, softened water, and protected both clothing and washing machines from daily wear and tear.

But phosphates, like other apparently benign materials such as asbestos or chlorofluorocarbons (CFCs), have undesirable environmental side effects. By the end of the 1960s, there was increasing concern about the impact that phosphates used in detergents were having on waterways. The problem stemmed from the fact that phosphates are a type of fertilizer; when released with the water after washing, they can stimulate an excessive growth in algae in slow-moving rivers and lakes, which use up oxygen and kill other living things. This process is called eutrophication. Mounting scientific evidence linking phosphates with eutrophication caused understandable concern in the detergents industry. The German detergents producer HENKEL decided to invest in the search for an effective substitute. This program resulted in not only the development of Sasil, currently the world's leading phosphate substitute, but also the successful phase-in of phosphate-free detergents in Germany and other European markets.

HENKEL is a German-based family-controlled consumer products and specialty chemicals company with worldwide sales of more than DM12 billion ($7.6 billion) in 1990. Its extensive range of detergents and household cleaners—including Persil, introduced in 1907—account for almost a third of total sales. The environmental sensitivity and high public profile of its products are a result of HENKEL's strong stand on environmental issues early on. For example, Konrad Henkel, chief executive officer in 1972, declared that "one cannot simply sell one's products without assuming responsibility for them."

Work on a phosphate substitute started in the late 1960s under B. Werdelmann, then a member of the board of directors. At the height of the development process more than 200 HENKEL employees were working full-time on the project. The alternative had to be just as effective as phosphates at softening water and removing dirt. It had to be odorless and have an acceptable consistency. But it also had to be tested for environmental compatibility to ensure that solving one environmental problem did not result in another. In 1973, after more than a decade of intensive research and development, costing approximately DM120 million ($76.5 million), HENKEL found the favored phosphate substitute, Sasil. The fact that Sasil was the first detergent additive created at HENKEL that was not water-soluble led to some friction within the company, with many experts joking that the Sasil team was "planning to add sand to HENKEL's detergents." But by demonstrating the effectiveness and environmental benefits of the new additive, the team managed to win over internal critics.

The simple discovery of Sasil was not the end of the story: the new product had to be tested for consumer and environmental acceptability. Many feared that the use of Sasil would lead to a buildup of sludge in sewers and surface waters in the area. So in 1976 HENKEL conducted large-scale consumer testing with more than 1,000 households in the Baden-Würtemberg region. The company also undertook an extensive environmental impact assessment of Sasil with the University of Stuttgart and 15 other public and private institutions. The tests showed no negative side effects of Sasil on the environment.

The actual introduction of Sasil-based detergent to the wider market was influenced by two factors: legislative pressures and consumer demand. In the 1970s, states bordering the Great Lakes in the United States had passed laws limiting the percentage of phosphates in detergents. The Federal Government of Germany also took an active part in the effort to develop a phosphate substitute, but was careful not to pass legislation limiting the use of phosphates before an acceptable substitute was available. In 1980, the government acted to accelerate the process by issuing a decree that required the step-by-step removal of phosphates from detergents, so that by 1984 detergents in Germany could contain no more than 20 percent phosphates. HENKEL had met this requirement by 1981. By the end of the 1970s, market researchers were also showing increasing demand among German consumers for phosphate-free detergents. Yet many consumers still believed that detergents with phosphates were more effective than the alternatives.

By 1982 enough evidence had been gathered that Sasil was environmentally safe and production capacities for Sasil had been expanded sufficiently to remove the phosphates from Dixan completely and replace them with Sasil. The company decided not to present the product as phosphate-free in order to test consumer perception of products containing Sasil. Consumer acceptance of Dixan remained unchanged. This success convinced HENKEL marketing personnel that a relaunch of its market leader, Persil, with a phosphate-free formula could be successful in both Germany and Switzerland.

In 1986, Jürgen Seidler, director of detergent marketing in Germany, and Jörn Jobs, product group leader for heavy-duty detergents, developed a two-pronged strategy. They decided to launch a phosphate-free version (which would be aimed at ecologically oriented consumers) while retaining the traditional blend. As Jobs said, "It was clear by 1982 that Sasil was an effective and safe substitute for phosphates....Our next job was convincing our customers of the fact."

The central theme of the marketing strategy was "Persil remains Persil." To reinforce the continuity of the brand, the packaging was not changed, except for the addition of "phosphate-free" to the front of the box. Despite somewhat higher production costs for Persil phosphate-free, both versions were sold at the same price. The two brands were also advertised together in all television, radio, and print media to stress the point that the consumer had a choice. To get the message across, HENKEL raised Persil's advertising budget from DM20.1 million to DM25.4 million ($12.8 million to $16.4 million).

The strategy worked. Within a year of the launch of the phosphate-free Persil, sales had risen significantly. Shortly after its introduction in Germany, phosphate-free Persil was launched in Austria, Belgium, the Netherlands, and Switzerland. Due to overwhelming consumer acceptance of the new Persil, all HENKEL's major competitors launched phosphate-free detergents within a few months. Sales of phosphate-free Persil steadily increased until, by January 1989, the old phosphate version was phased out. Nonphosphate powder detergents now have a 100-percent market share in Austria, Germany, Italy, the Netherlands, Norway, and Switzerland. The same situation applies in Japan, while in the United States, Congress has proposed legislation that would require all detergents sold there to be phosphate-free by 1996.

HENKEL has recognized that even within Europe, consumer tastes and expectations differ, greatly influencing the spread of phosphate-free detergents. "All markets in Europe are not exactly like the German market," says Klaus Morwind, head of HENKEL's detergent business. HENKEL tailors its products based on consumer expectations, market conditions, and, more specifically, legislation governing the use of phosphates in detergents. By the beginning of 1991, phosphate-free powdered detergents had 60 percent of the market share in Belgium, 70 percent in Finland, and between 30 and 60 percent in Denmark, France, and Sweden.

The introduction of phosphate-free detergents has also not been without resistance from competitors. Clearly, it challenges those companies that have not been farsighted enough to develop environmentally acceptable substitutes. In Germany, a number of phosphate manufacturers that paid less attention to research in phosphate substitutes have been forced to lay off workers and close production facilities. In France, the leading phosphate producer, Rhône-Poulenc, launched a counterattack when HENKEL introduced the phosphate-free detergent Le Chat in the spring of 1989. Riding on the wave of increasing public concern about the environment, sales of phosphate-free powders in France tripled, largely due to Le Chat. Rhône-Poulenc claimed in its advertising campaign that phosphate-free detergents were both ineffective and more harmful for the environment. The strategy backfired, however, and Rhône-Poulenc was forced to withdraw the advertisements. Furthermore, the French government has recently passed a new law that limits the amount of phosphates in detergents to 20 percent. Le Chat now has a 4-percent market share.

The results of the phaseout of phosphates in detergents has been impressive from both an ecological and a commercial point of view for HENKEL. In Germany, the amount of phosphates in laundry detergents dropped from 275,000 metric tons in 1978 to zero in 1991. The decision to invest in the development and production of Sasil paid off for HENKEL. HENKEL is receiving worldwide royalties—except from the United States—from Sasil licenses, and Persil's market share was enhanced. The benefits, however, cannot be measured purely in financial terms.

HENKEL's public image has been bolstered, coming second behind Volkswagen in a 1986 opinion poll in which Germans were asked which company they thought did most for the environment.

Lessons Learned

• Companies need to have vision and foresight to tackle the environmental challenge. They must be ready to change well-established product lines to meet new consumer demands, legislative requirements, and scientific discoveries.

• Environmentally driven innovations need a champion at the highest level.

• The introduction of new, more "environmentally friendly" products has to be handled with care to convince consumers that they are not sacrificing product quality for environmental safety.

Contact Persons

D. Barker, E. Smulders, and
 B. Schmidt
HENKEL KGaA
Henkelstrasse 67
Postfach 1100
D-4000 Düsseldorf 1, Germany

Case 16.5
Laing: Energy-Efficient Housing

Domestic housing consumes a substantial amount of energy for heating, lighting, cooking, and other purposes. In areas such as Scandinavia, North America, and Australasia, energy-efficient timber frame houses are already well established, forming the majority of houses built. In the United Kingdom, however, the vast majority of houses are built using a traditional wall construction of exterior brick, cavity, and then internal concrete block. Laing Homes, a division of John Laing plc, realized that introducing timber frame houses into the U.K. market would bring energy savings to its customers and financial returns for itself, and so in 1979 it started to build such homes with a high level of insulation.

According to Pauline Land of Laing Homes, the driving force behind the development and introduction of timber frame houses was profitability: "The new product was more profitable and also fulfilled many of the Company's environmental criteria." After reviewing the building methods available in 1979, Laing realized that they were able to achieve certain commercial advantages by adopting a new system. The walls of the timber frame houses could be prefabricated and distributed in containers from warehouses, saving on construction costs and site traffic. Construction time was reduced, which meant that in a large development, units could be built as they were sold.

By filling the air pockets in the timber frame with mineral wool insulation, Laing also managed to cut the rate of heat loss—the U-value—to 0.37 watts per square meter degrees Kelvin (W/m^2K), substantially below both the existing 1.0 W/m^2K rate required by U.K. Building Regulations for new properties, and the tighter standard of 0.6 W/m^2K that was introduced in 1982. Using the National Homes Energy Rating computer program, this means that a Laing timber frame design brings annual energy savings of 12 percent, compared with a similar three-bedroom detached house meeting the 1982 regulation. This saving continues each year for the life of the dwelling, which is expected to be more than 100 years. The timber frame house also consumes 18 percent less energy in the manufacture and distribution of the construction materials. In other countries, the energy savings could be even higher, since most of the timber used in the U.K. construction industry is imported from areas such as Scandinavia and Canada.

Despite having a product that brought significant savings to the consumer, Laing recognized that it could be extremely difficult to get both the public and the building industry in the United Kingdom to accept the timber frame construction. Public perception favored something traditional and "solid," and timber frame in the public's opinion did not provide this. To counter this belief, Laing developed a marketing strategy emphasizing the intrinsically low energy consumption of the Laing timber-framed dwelling. Typically, the display boards in sales offices stress the energy efficiencies in the house. This marketing strategy readily fitted into the Laing ethos, as the company has for many years led environmental initiatives within the construction industry. However, conventional marketing wisdom says that energy is not high on the priority list when the average consumer looks at a new house, and the marketing strategy was severely tested in the mid-1980s.

In 1984, a television documentary criticized timber frame construction in the United Kingdom, concentrating on condensation and the use of damp-proof membranes within the construction. In reality, the number of houses with problems was extremely small, but almost overnight the timber frame market disappeared. Companies building many times the number of timber frame houses than were being built by Laing Homes decided to revert to traditional U.K. construction techniques. Laing Homes, however, was almost totally unaffected; the only real difference was that Laing marketing featured the energy efficiency characteristics of timber frame homes. In addition, a very strict quality control program had been implemented.

Timber frame walls require different construction methods, and Laing Homes had to develop specially tailored design techniques, training, and quality control procedures. Of these, the training of subcontractors was perhaps the most difficult, but the company has now developed a comprehensive construction manual for its subcontractors, which provides the basis of all their work. For any major new project, a method statement is issued detailing the construction procedures. All sites are subject to regular quality assurance inspection by Laing Homes head office. The results have been impressive, with a very low incidence of return maintenance visits.

Laing Homes sold 2,000 timber frame houses in 1990, which represents 75 percent of its total construction (the other 25 percent is accounted for by social housing projects that stipulate "brick and block"). As the timber frame range became established within Laing, other improvements were sought. Environmentally sound practices were being implemented throughout the company during the 1980s, and Laing Homes required that suppliers conform to the group's environmental policies. For example, hardwoods (where used) have to originate from sustainably managed forests, and CFC-blown insulants were replaced with CFC-free materials. Replacing CFC-based materials used to insulate hot-water storage cylinders has been more problematic, as the alternatives are more expensive and lead to an increased heat loss from the cylinder.

Laing's persistence with its timber frame houses has paid off, and in 1991 it was named the House Builder of the Year, the major consumer-judged award in the country, based on quality of product and service. But the company recognizes there are still many things that can be improved. It has developed ways of achieving much lower U-values, and is considering whether to develop "super energy-efficient" houses for niche markets in the future. Certain design features also need to be tackled. For example, improvements could still be made to maximize the amount of sunlight trapped by the house through changes to the glazing area and the orientation of the site.

By implementing all realistic measures, it may well be possible to save up to 20 percent more on the energy required to build and run a typical house using today's technology. Land explains that "the main reason these haven't been developed is competition." The long-term decline in real energy prices during the 1980s and the short-term building slump in the United Kingdom mean that developing further energy-saving initiatives could undermine Laing Home's current competitive edge. To overcome these obstacles, Laing believes there needs to be a continuing rise in housing energy efficiency standards, as well as the use of proper market signals to promote the construction of more energy-efficient housing and the retrofitting of existing homes.

Lessons Learned

• Designing and providing energy-efficient products can be combined with corporate profitability.

• Companies need to inform and convince customers about the advantages of energy efficiency.

Contact Person

P.K. Rees
Group Director
Marketing and Information Services
John Laing plc
Page Street
London NW7 2ER, UK

Case 16.6
Volkswagen: Recycling the Car

In the not-too-distant future, new cars could leave the factory gate passing old cars returning as recycled materials to be used once again in the production process. To some extent, this remanufacturing cycle is already in operation for steel, which makes up more than 75 percent of each car. As Rolf Buchheim, manager of research and development at Volkswagen (VW), observes, "no one would dream of asking whether the steel for the mudguard was used previously as, say, a bonnet."

The problem, however, is with the remaining 25 percent—approximately a quarter of a metric ton of material per car—which is currently landfilled. Most countries in Western Europe and North America are fast approaching the social and environmental limits to landfilling. Furthermore, landfill represents an avoidable waste of useful materials. As a result, auto manufacturers are now rethinking the design and stewardship of their products to achieve as near as possible 100-percent recycling. The conceptual and operational challenges are immense. Cars have traditionally been designed with ease of design and use in mind; now ease of disassembly and reuse has to be factored in.

Taking a life-cycle approach puts the disposal phase in perspective. Across the entire life of a car, almost 90 percent of energy consumption occurs during its use. Although this energy cannot be recovered, it can be reduced. One of the best ways of promoting energy conservation is through reducing auto weight, replacing steel with plastics (or aluminum). But plastic wastes—which now account for over 10 percent of car weight—are not yet managed in a systematic way, restricting the possibilities for increased lightweighting.

In Europe, German manufacturers are in the vanguard. Environment Minister Klaus Töpfer has threatened to force automakers to take back the 2 million cars that are scrapped in Germany each year if they cannot arrange an effective recycling system by 1993. With this narrow window of opportunity, Germany's automakers are racing to establish the most efficient disassembly systems. Among them is Volkswagen, Europe's largest automaker.

At Volkswagen, only 6 percent of the total production material is removed as waste; 62 percent goes into the product and 32 percent is now recycled. VW has already largely closed the plastics loop at the production stage. As a result of VW's efforts at its production plants, nearly 100 percent of plastic waste is recycled. The differing polymers are separated, and remolded for the same purposes. More than 70 different components are currently made from recycled materials. The focus now is on developing systems to maximize the recycling of plastics waste at the end of the product life cycle.

A hierarchy of resource conservation options has been established to guide decision making: long-life product, reutilization of used parts, recycled materials used for original purposes, recycled materials of inferior quality, thermal recycling, landfill site, and—last—hazardous waste site.

According to Buchheim, although the first two options—designing long-life products and reutilizing used parts—are superficially attractive, they can stand in the way of technical progress. For example, the drag coefficient for cars improved by more than 30 percent between 1970 and 1990. By keeping cars on the road longer, the introduction of newer, more-efficient models could be constrained, he argues. As a result, Volkswagen's aim is to channel as much as possible from a scrapped car back into the production cycle; "if possible, 100 percent recycling," says the company. This is a complex issue: each car has thousands of components, made from a variety of different materials, metals, glass, plastics, and rubber. Therefore to recycle a scrap car, several different resource loops need to be established.

Most of the metals and some of the rubber have long been recycled. For plastics, top priority goes to reusing materials at the same quality level. "Our objective must not be to recover valuable automotive plastics and then turn them into plastics of inferior quality for use in park benches, flower pots, or sound-proofing walls," argues Buchheim. Not only must the resource loop be closed to avoid emissions to the environment, it must also be closed to maintain the value of the various waste streams. VW estimates that, for example, 70 percent of the plastics used in the VW Passat are currently recyclable. The issue was to find out how the company could realize this potential in practice and push it closer to the 100-percent goal.

Since the beginning of 1990, VW has been running a pilot auto recycling plant at Leer in northern Germany in collaboration with the local chamber of commerce and employment office, and together with a leading industrial raw materials company. By the end of 1991, more than 2,000 cars had been processed, giving VW invaluable experience in how best to both dismantle cars and design future ones to facilitate recycling.

At Leer, each car goes through a careful dismantling process, separating the various waste streams. First, all oils and other liquids are drained off; the company quickly realized that items such as closed gear boxes hampered this process. Batteries are removed so that the metals, plastics, and acid can be recovered. Plastic parts are removed, reground, and then transported to suppliers. Catalytic converters, which contain valuable precious metals, are also separated and processed. Finally, after the glass and rubber have been extracted, the remainder is shredded for use in the steel industry.

Success has already been achieved with recycling plastics. At Leer, plastic bumpers are broken down into their separate polymers at the plant. These materials are then returned to VW's supplier of bumper plastics, where they are recompounded and directed back into the production process for new bumpers. Already 20–30 percent of the bumper of the new VW Polo is made from recycled plastics; VW's ability to reach 100 percent is limited by logistics—that is, its ability to ensure sufficient supply of old bumpers for reuse.

VW's experience at Leer has shown that no longer is the cost of production the sole consideration in design and materials choices; "the costs of reuse must also be included," says Buchheim. This means reassessing existing designs and materials. Some broad lessons have already been learned. Simplification is a great aid to recycling. All major plastic parts are now marked for easy recognition during disassembly; clips are favored instead of screws, and the number of different components and materials used for a particular module of the car is minimized where possible. There have already been some design improvements: for example, the fuel tank of the old Golf model had 32 different parts, made from several different types of plastics. The fuel tank for the new model Golf contains only 16 parts in an integrated design. Disassembly time has been cut from seven to three minutes. Even small parts like engine bearings have been redesigned to facilitate recycling.

In addition to better design, VW hopes to improve its recycling rate through choosing materials that can operate within a closed-loop system. Materials recyclability is now included as one of the design goals for each new model at VW, and "as a matter of principle, new materials are only given the go-ahead for full-scale production if they are integrated in a self-contained recycling concept." Clearly, some plastics will gain market share at the expense of others.

VW has developed a model of how a future recycling system could work. Cars would be returned to a decentralized network of licensed recyclers, who would break them down into component parts and then sell the materials back to the relevant industries—steel, plastics, glass, and so on. "It's not economic to transport old cars over long distances," believes Buchheim. The supplier industries would then process the materials and feed them back into the automotive and other production cycles as appropriate, "reintroducing these high-quality recycled materials into their primary cycles." VW will also include a preference for high-quality recycled materials in its supplier contracts to facilitate this process. The knowledge gained at Leer will be transferred to VW-approved recyclers. The German government will decide on its proposed system for auto recycling during 1992, and looks likely to approve VW's concept of licensed recyclers.

VW's pioneering work on car recycling at Leer has enabled it to announce that it will take back the new model Golf for recycling free of charge. However, because of the average 10-year car life span, Golfs will not come back in significant numbers until after 2000. Nevertheless, this commitment was one of the factors that led to the new Golf being given the 1992 European Car of the Year award.

Lessons Learned

• Given an appropriate design and a complete logistics system, even complex products such as cars can be recycled.

• Environmental innovations can boost corporate image and product sales.

• Pioneering work by companies can help influence future government regulations.

Contact Person

Rolf Buchheim
Manager, R&D Volkswagen
W-3180 Wolfsburg, Germany

Case 16.7
Pick'n Pay: Retailers and Sustainable Development

Retailers are where industry and consumers meet. Retailers are the first to hear the complaints and wishes of individuals. They are also in a position to educate the consumer and to influence suppliers about products to be offered for sale. As the public has become more sensitive to environmental concerns, many retailers have realized that attention to environmental details can provide opportunities to capture increased market share. Providing customers with accurate, understandable information, with choices, and with effective products having good environmental characteristics is good for business and good for nature.

South Africa is a Third World country, albeit with First World components. But the country is now experiencing a "silent revolution" as underlying social and economic forces transform its society in a process as dramatic locally as the Industrial Revolution was in Britain.

This revolution is one in which black South Africans have increased their share of the country's disposable income from 32 percent in 1970 to nearly 50 percent today. Higher incomes among blacks will promote economic development. However, as Wendy Ackerman, Pick'n Pay director, says, "environmental initiatives in the Third World will stumble unless supported by sustained economic growth and social justice." As one of South Africa's largest retailing chains, Pick'n Pay has decided to use its crucial gatekeeper role to promote sustainable development.

The start of Pick'n Pay's initiative was an internal environmental audit, which revealed a number of areas of concern, including packaging and waste disposal, environmentally harmful products, energy profligacy, and the use of CFCs in refrigeration and air conditioning units. The audit also pinpointed the fact that staff were insufficiently aware of the company's environmental policy and the principles on which it was based. As a result of the audit, Pick'n Pay published an "environmental mission" in all the major national media throughout South Africa, one of the first detailed public commitments to environmental excellence made by a local corporation. The mission committed the company to take action on 14 key issues, ranging from improving site operation to choosing more sustainable produce and contributing to society's environmental education.

The company also launched a line of environmental products with its Green Range, including natural toiletries, phosphate-free detergents, and mercury-free batteries. Furthermore, the packaging for the Green Range is scrutinized to ensure that all cardboard and labels are made from recycled materials and that any plastic packaging is made from recyclable polyethylene. Pick'n Pay has worked closely with 15 suppliers to develop the Green Range to facilitate the changeover of manufacturing systems and introduction of new product ingredients. Green Range products are beginning to move producers and consumers toward more environmentally friendly products.

Within the company's stores, Pick'n Pay has introduced computerized electrical management systems to conserve resources. Savings are also made in new buildings by incorporating as much natural light as possible in the design. Containers for plastic, glass, and tinplate have been placed at all stores. The company has also replaced CFCs in its refrigeration units with HCFC-22, the coolant with the lowest ozone-depletion potential available in South Africa. Pick'n Pay is encouraging local farmers to produce organically grown fruit and vegetables and to minimize the use of pesticides. The company conducts spot surveys to confirm farmer cooperation. An innovative computer-based packaging analysis scheme to assist buyers in making an informed choice where manufacturers' products are concerned is being evaluated based on a system developed by Migros in Switzerland (see case 16.3). Selection of a particular supplier's product is therefore based on the dual criteria of product and packaging ingredients.

Promoting awareness and commitment within the company has been central to the environmental mission. Lack of "environmental literacy" among staff presented an initial stumbling block. Environmental issues have been incorporated into the company's mainstream staff training program. Employees have also formed "Green Groups" dedicated to sponsoring environmental projects in domestic and work communities.

Pick'n Pay has also been active outside its stores. According to Pat Irwin, professor of environmental sciences at Rhodes University, the best remedy for public apathy toward environmental issues is "collective change in attitude toward education in general and environmental education in particular." Based on this principle, Pick'n Pay launched a nationwide environmental education program in October 1991 in cooperation with the local World Wide Fund for Nature affiliate, the Southern African Nature Foundation. In addition, more than 4 million "Enviro Fact Sheets" were distributed free to children. The company has sponsored numerous competitions, cleanup drives, and environmental projects run in conjunction with local schools, involving more than 1 million children. Special projects have encouraged members of the public as well as employees to plant hundreds of trees in their local communities. The company has also lobbied government to produce lead-free gasoline since 1988.

Lack of public awareness of the company's environmental policies were the cause of some skepticism early on that Pick'n Pay's initiative was driven by marketing considerations. As the policy has been implemented, however, public perception has undergone positive change during the past two years, and the company's environmental credibility was further enhanced when it won the corporate category of the Southern African Nature Foundation's prestigious national environmental award in June 1991.

Pick'n Pay has formulated a code of ethics, or corporate environmental policy, approved at board level. The chief executive officer has publicly and regularly stated his involvement in and commitment to the concept and applicability of sustainable development in South Africa's Third World environment. As Pick'n Pay chief executive officer Ray Ackerman says, "Clearly environmental thinking has gone way beyond wildlife conservation and industrial pollution, and any retailer who ignores this does so at his peril."

The company takes a global vision of environmental management, leading it to adopt a cooperative rather than a confrontational approach to environmental stakeholders. "Sustainable business is a very practical concept. In my view, it is one of the most crucial issues of this decade. We cannot begin to address the massive problem of poverty without it," says Ackerman. "It is important to realize that a good company reputation buys credit—not simply in financial terms, but with the local community and therefore with its market base."

Lessons Learned

• Retailers in industrial and developing countries can provide a public service by being leaders in green consumerism.

• Environmental awareness should become an integral part of a company's annual business plan. Good business begins when responsible business leaders bring environmental concerns into the professional decisions they make during their working hours.

Contact Persons

Wendy Ackerman
Raymond D. Ackerman
Pick'n Pay Stores Limited
Pick'n Pay Centre
Corner of Main & Campground Roads
P.O. Box 23087
Claremont 7735, South Africa

Case 16.8
ENI: Developing a Replacement for Lead in Gasoline

In 1968, the ENI Laboratories synthesized a new chemical product called
methyl tertiary butyl ether (MTBE). This product was developed as part
of a research effort to find a commercial use for a waste by-product.
Initially MTBE was marketed solely as a solvent; ENI researchers discovered, however, that MTBE was an efficient octane enhancer that could be
used as an alternative to the "lead" additives then widely used in
gasolines.

There are two reasons for getting the lead out of gasoline. First, lead
is toxic and can harm the health of those exposed to high levels of it.
Children are particularly susceptible to some of these effects. Second,
lead can "poison" the platinum in the catalyzers used on new automobiles to control air pollution emissions. Catalyst-equipped cars require
lead-free fuel with high octane levels to ensure good engine performance. Vehicle emissions have historically been one source of lead in the
environment.

In 1973, ENI brought on-line the world's first production facility, at
Ravenna, Italy, to make MTBE for sale as a solvent. But the company
recognized that MTBE could reach an even wider market as an octane
enhancer and quickly began selling the product to refiners in Germany.
The market for MTBE increased rapidly as programs to limit the lead
content in gasoline were introduced. The production of unleaded gasoline began in the United States in the late 1970s, creating a market for
MTBE and other non-lead octane enhancers. In Europe, as well as in a
number of countries outside it, the phaseout of lead in gasoline is
expected to be completed in this decade.

Unleaded gasoline contains between 10 and 15 percent MTBE, and demand continues to grow. It is a case where a decision in favor of the environment opened up an interesting new business area.

MTBE was traditionally manufactured in petrochemical plants and refineries by using the by-product iso-butylene from steam crackers and catalytic crackers. MTBE production was therefore connected to the operating capacity of those plants and thus limited by the availability of this key raw material. Many small-scale MTBE plants were built according to that scheme, and total world production was only 1 million tons in 1980, of which two thirds was used in the United States and one third in Europe.

In order to meet future demand (a complex forecasting problem), ENI sought a second technological breakthrough—the use of isobutane dehydrogenation technology to produce larger quantities of the basic raw material, iso-butylene, thereby providing adequate supplies of the feedstock to make "world-scale" MTBE plants feasible.

This technology was proposed to Saudi Arabia, where Ecofuel (a subsidiary of ENI) established a joint venture company with SABIC—the Saudi petrochemical corporation—and with other partners. The new technology was used for the construction of a plant capable of producing 500,000 metric tons of MTBE annually. It began operating in 1988, the first of that large-scale capacity to use the Ecofuel "dehydrogenation" technology.

The same experience was applied to a second joint venture, incorporated in Venezuela between Ecofuel and the Venezuelan company Pequiven, to build another world-scale MTBE plant that began operation in 1991.

MTBE from these plants, marketed largely by Ecofuel, is used worldwide in gasoline distributed in Italy and Europe through the Agip Petrol's distribution network and sold to international customers requiring this octane-enhancing component for their gasoline blends. The production system identified by ENI and by a small number of other pioneers to produce gasoline using MTBE is now well developed throughout the world.

The demand for MTBE in 1990 had increased to about 8 million metric tons, and is expected to increase to about 15 million tons in 1995 and to more than 20 million by the end of the decade. New programs to control smog created by photochemical oxidants will probably require further reductions in hydrocarbon emissions from cars. Even though today's technology eliminates 95 percent of all hydrocarbon emissions from a new 1992 automobile, cleaner fuels with more "oxygenates" like MTBE may be needed in the future.

As a result, ENI intends to remain one of the major participants in MTBE technology and production. ENI has recognized that working to improve the environment is good for its public image and good for the bottom line as well.

ENI is planning new investments in the MTBE sector to maintain its market share and is studying intensively another technological jump to make MTBE even more competitive in the market. The company wants to be ready to provide the market with ecological fuels for the next generation of energy products.

Certain cities or regions of the world with unique pollution problems have already introduced more stringent regulations concerning car emissions to be implemented in the next several years. These requirements will further limit polluting agents such as sulfuric anhydride, benzene, carbon monoxide, nitrogen oxides, and ozone-forming chemicals. As new regulations appear in more and more areas, the challenge for industry is to be prepared, through advanced research and investments, to provide consumers with the clean fuels they need.

Lessons Learned

• Industries can realize business opportunities by finding new uses for waste products.

• Research can help industries make profitable investments in environment-oriented activities and cleaner products to meet demands.

• Positive economic and environmental results can often be obtained through joint ventures in developing countries with available raw materials, such as natural gas.

Contact Person

Francesco Cima
General Manager for Development
Ecofuel
Viale Brenta 15
I-20139 Milan, Italy

17 Managing Sustainable Resource Use

The use of resources for energy production, for increased agricultural activity, and for forestry and land management is crucial to economic growth and development. Without energy, we cannot produce the goods and services to meet human needs. Yet energy production and use often have negative environmental impacts. Without productive agriculture, the world's increasing population cannot be fed adequately. Yet certain agricultural systems degrade the productivity of land and pollute scarce water resources. Forests and lands provide wood and habitat for a wide variety of animal and plant species, and serve as natural air and water purification systems. Yet careless use of forests or land can unnecessarily destroy this resource base and deprive future generations of these services.

This chapter looks at five cases where companies and individuals are beginning to manage resources for energy, agriculture, and forestry purposes in a sustainable manner.

Case 17.1 describes how coal can be used more cleanly to produce less pollution per unit of usable energy and more efficiently to use less coal per unit. Using coal more efficiently reduces the amount of carbon dioxide—a greenhouse gas—building up in the atmosphere.

In Zimbabwe, Triangle Limited converts renewable plant material, biomass, into a substitute for expensive imported petroleum products, as case 17.2 describes. In India, the sugarcane processor E.I.D. Parry has introduced better agricultural processes to make farmers and its sugar plant more profitable and less polluting, as case 17.3 describes.

Case 17.4 explains how the Brazilian company Aracruz replanted a large deforested area and now produces more than 1 million tons of pulp on a renewable basis. In Australia, ALCOA has an award-winning program to protect the endangered jarrah forests and to reclaim land after mining, discussed in case 17.5.

Case 17.1
ABB: Introducing Clean Coal at Värtan

Developing countries require a resurgence of clean and equitable economic development to tackle the challenges of mounting poverty, population growth, and environmental degradation. The question of how to
fuel this development is critical for the long-term sustainability of both
North and South. Emissions of carbon from the burning of fossil fuels are
responsible for a considerable portion of greenhouse gases associated
with global warming. Coal is the fossil fuel with the most carbon per unit
of energy.

Significant reductions in carbon emissions could be realized if the
world switched from coal to a fuel with less carbon—something easier
said than done. China and India, for example, together only have 2
percent of natural gas and oil resources, but they have 20 percent of world
coal resources, which they intend to exploit to promote much-needed
economic development.[1] If China alone realizes its plans to triple its coal
consumption by the year 2030 using existing technologies, global carbon
emissions from coal could increase by 50 percent. Given the prospect of
increased coal use to meet development needs, the world needs clean
coal technology to keep traditional pollutants out of the air and efficient
technology to minimize output of carbon dioxide per unit of energy.

Considerable "clean coal" technology already exists, and more is being
developed. Among the most promising technologies are integrated coal
gasification combined cycle (IGCC) and pressurized fluidized bed combustion (PFBC). With IGCC, more than 99 percent of the sulfur can be
removed before combustion, and nitrogen oxide (NO_x) levels are similar
to those of natural gas combustion. Through the use of both a gas turbine
and a steam turbine in a combined cycle, efficiencies exceed 40 percent,
with a potential for further improvements. Several companies are involved in the development of this technology.

PFBC involves mixing crushed coal with limestone and burning the
mixture in jets of air. Efficiencies can be raised to 40–44 percent today
through a combined cycle, with a potential for improvements to 50
percent. Compared with the average plant in developing countries, with
only 25 percent efficiency in converting coal into electricity, an advanced
coal plant with, for example, 43 percent efficiency saves more than 40
percent in coal and carbon emissions; other emissions would be cut by

90 percent.[2] When designed to generate both heat and power, the total efficiency is about 90 percent. The produced heat can, for example, be used in a district heating system. District heating also has the advantage of increasing energy flexibility. Heating systems can be gradually expanded, and switching to sustainable energy sources in the future would be facilitated.

In the mid-1980s, Stockholm Energi, a public utility in Stockholm, was considering different means of generating additional electricity and heat. The plant would have to meet stringent efficiency and pollution control requirements. The Swedish government had pledged to cap the country's carbon emissions as part of its contribution to limiting greenhouse gases and wanted to reduce dependence on imported oil. A strong environmental movement had meant that further expansion of nuclear and hydroelectric capacity was out of the question. As a result, Stockholm Energi chose to construct a new coal-fired power plant. The issue was whether to choose a conventional plant with flue gas desulfurization pollution control, an IGCC plant, or a PFBC plant. Among the criteria used for the selection were efficiency, emissions, compactness, and cost; compactness was needed because the Värtan site was in the center of the city. The only technology that could meet all these requirements was ABB Carbon's PFBC. Construction began in 1987; four years later, the combined heat and power plant was connected to the grid.

The plant consists of two PFBC modules, producing 135 megawatts (MW) of electricity and 224 MW of heat, with a total efficiency of 89 percent. The heat is used in the local district heating network, reaching more than 70,000 households. Out of Stockholm's total energy consumption, as much as 60 percent is for heating. While the decision to install a coal-fired power plant in the city initially upset environmentalists, nearby residents, and the neighboring city of Lidingö, Värtan is gradually being accepted by all groups because of its high performance and low environmental impacts.

The Värtan plant has demonstrated that coal technologies can meet very strict emission standards: yearly averages of 30 milligrams per megajoule (mg/MJ) for sulfur and 50 mg/MJ for NO_x, and a monthly average of 10 mg/MJ for particulates. At full load, NO_x emissions have been as low as 20 mg/MJ and particulates below 1 mg/MJ. Sulfur emissions are reduced by adjusting the limestone injection rate, so that they comply with government regulations. This means that the plant safely meets government emission standards. In comparison, an oil-fired plant emits 240 mg sulfur per MJ when burning 1-percent sulfur oil. The average NO_x emissions of all oil-fired plants at the Värtan site is 180 mg/MJ.

The price of the Värtan PFBC plant was rather high due to permit requirements and site location. According to Stockholm Energi, the total expenditure was Skr1.3 billion (1985 krona) ($230 million). Because of the low number of commercial PFBC and IGCC plants, it is difficult to compare the costs of "clean coal" with traditional coal technologies. However, an evaluation of PFBC technology made by the Bechtel Group in 1991 estimated that the cost of a new PFBC electric plant would be comparable to a pulverized coal plant with flue gas desulfurization (FGD).[3] A study by the U.S. Department of Energy gives second-generation PFBC and IGCC plants significant cost advantages over conventional coal with FGD.[4] According to Krishna Pillai, vice president of ABB Carbon, in countries such as Sweden where standards for sulfur emissions would require the installation of desulfurization equipment, PFBC is already competitive with conventional systems.

Pillai admits, however, that PFBC will not be able to compete against conventional technology in countries lacking emission regulations. In the face of the global imperative to reduce carbon dioxide and other forms of pollution caused by the burning of fossil fuels, the issue is how this "clean coal" technology can be made available to developing countries that need it but cannot pay for it. A first step would be to ensure that existing sources of development finance from institutions such as the World Bank specify "clean coal" technologies in their energy supply projects. For multiple investments, another way of reducing costs of importing equipment is to produce the power plants locally through joint venture or licensing agreements.

Lessons Learned

• Clean coal technology is a cost-effective way to generate electricity in a more environmentally compatible manner.

• In view of the greater capital requirements, innovative financial arrangements will be needed to spread this clean technology throughout the developing world.

Contact Persons

Jan Sjödin	Ove Gustavsson
Plant Manager	Marketing Manager
Stockholm Energi AB	Sales and Development
S-113 91 Stockholm	ABB Carbon AB
Sweden	S-612 82 Finspång, Sweden

Case 17.2
Triangle Limited: Energy from Biomass

Biomass in the form of fuelwood is the primary energy source for almost half the world. Energy can also be generated from plant matter to produce ethanol, a renewable and cleaner alternative to petroleum for transportation needs. The production of ethanol can contribute to greater self-reliance in developing countries by reducing dependence on imported oil. In Zimbabwe, a project started under the previous minority government to generate ethanol from sugarcane to cope with international sanctions has been continued following the introduction of majority rule as a way of coping with excess sugar production and reducing oil imports.

In 1978 the minority Rhodesian government was the subject of international economic sanctions and was increasingly concerned about its ability to obtain gasoline. The country's landlocked position, the vulnerability of its supply routes, foreign-exchange limitations, and strategic considerations were all major factors in the decision to develop alternative fuels from domestic renewable resources. The country has a strong agricultural base that normally produces a food surplus. Therefore, it was unlikely that "food versus fuel" would be used as an argument against using biomass to produce fuel. Sugarcane was a major crop, particularly in the southeastern low veld areas near the towns of Triangle and Chiredzi.

Triangle Limited was a major sugar producer in the region, using sugarcane grown on the two largest estates in the country. The export price of sugar was poor, close to or less than the cost of production, so the conversion of sugarcane to a fuel supply made economic as well as strategic sense. The sugar refining by-product, molasses (uncrystallizable sugars and solubles), had also been in surplus supply for many years. As a result, Triangle was pleased to develop plans to use a portion of its sugarcane and surplus by-product to produce ethanol, a substitute for gasoline. The final proposal for a plant that could produce 40 million liters per year was submitted and approved in 1978, and construction started in 1979.

The plant was designed by Gebr. Hermann, a German engineering company, which agreed to provide only the plans and technical supervision to ensure the maximum use of local materials and expertise. The equipment was a conventional batch fermentation process, a technology that could easily be learned and adapted to local operating conditions in Africa. All construction was to be carried out in Rhodesia. A special team was set up to translate the designers' specifications into locally available equipment and materials, cutting construction costs considerably and achieving a local content rate of 60 percent. At $6.4 million (1980 dollars), the Triangle plant was one of the most cost-effective in the world.

In 1980, the new government of independent Zimbabwe endorsed the project even though the economic sanctions had by then been lifted. The government was also concerned with balance-of-payments issues and wished to become less dependent on sugar exports, with their volatile prices and convertible foreign-currency earnings. The facility was designed to operate on a variety of feedstocks, thereby encouraging sugar production when market prices were high and ethanol production when world sugar prices were low. The ethanol plant was constructed as close as possible to the sugar mill to facilitate optimum energy use during production and the interchange of personnel.

To ensure a secure market for its ethanol, Triangle Limited negotiated a contract with the National Oil Corporation of Zimbabwe (NOCZIM) whereby the entire output of the facility would be purchased and resold to domestic oil companies for blending and distribution. The target blend ratio was 13 percent ethanol to 87 percent gasoline. Because ethanol is an octane-enhancer, NOCZIM would be able to import a less expensive nonpremium grade of gasoline. This had the added benefit of eliminating the lead content in gasoline. Although ethanol-sensitive plastic fuel lines in cars had to be replaced, no other engine adjustments were required.

The cost of producing the ethanol is 11–27 percent more than the price of gasoline delivered in Harare, but this does not include the foreign-exchange benefits or the contribution made to reducing excess sugar. Furthermore, the differential between gasoline and ethanol costs rises and falls based on the international market price for gasoline and sugar. Zimbabwe estimates that in 1984 the country saved $185 in foreign-exchange payments for each metric ton of sugar converted to ethanol (one metric ton of sugar can produce 590 liters of ethanol), for a total savings of $125 million. Of course, when the world market price of sugar is high, these foreign-exchange savings are reduced.

Since maximum resource efficiency is at the center of the existence of the Triangle plant, considerable effort has been made to ensure that as little as possible is wasted. For example, for every liter of ethanol produced, 9–10 liters of a hot (90 degrees Celsius) and acid (pH 4.5) liquid waste called stillage is generated. This material is high in potassium and phosphate, important fertilizer components. Triangle dilutes the stillage with irrigation water and disperses the mixture over 7,500 hectares of cane fields, thereby reducing the need to add potassium and phosphate fertilizers. There has been a 7-percent increase in cane yields as well as improvement in content of fermentable sugar in cane from fields treated with the diluted stillage waste. Water quality measured at 12 points within the estate and in adjacent streams is better than government standards and continued to improve during 10 years of operation. The filter cake or cane mud from the clarification process is recycled back onto fields as a soil additive.

Virtually all the sugarcane is put to use or recycled. After extracting usable sugar, the crushed cane (called bagasse) is used as the primary fuel source for energy for both the sugar mill and the ethanol plant. Historically in the sugar industry bagasse was regarded as a "nuisance" by-product. Surplus bagasse was burned simply to get rid of the material that could not be used in the refining process. However, the most recent boilers installed at Triangle have inlet-air heaters and water economizers that raise their efficiency to 82 percent compared with 55–60 percent for older units. Excess bagasse can now be diverted to animal feed. The boiler ash from bagasse combustion is also used around irrigation pipes to prevent erosion and aid water flow. During 1989, Triangle supplied 2 megawatts of electric power to the national utility, ZESA. Negotiations are now under way for Triangle to provide more electrical power to ZESA on a regular basis. Bagasse drying and storage would enable

Triangle to eliminate its use of coal during the two to three months when sugarcane is not available for processing.

There have been considerable development spin-offs from the project. The addition of the ethanol plant has increased the demand for sugarcane production in the entire region and provided more rural employment. Triangle has worked with local manufacturers to maximize local content of the ethanol plant and minimize reliance on imported spare parts. The company has established an on-site training school for welding and other technical skill positions. All new skilled workers are hired on a two-year training contract in special areas of expertise. A four-year training program was developed for electricians, boiler makers, and fitters. As a result, Triangle has acquired a permanent cadre of skilled workers in Zimbabwe. Preference in hiring and promotion is given to local nationals on the basis of merit.

According to Triangle, "cooperation between government and the private sector were crucial to the success of this project. It was essential that NOCZIM agreed to purchase and resell the ethanol to blenders and distributors. The government created market demand and the private sector produced the product to supply this demand." The plant has now been in operation for more than 10 years. The company is profitable and the nation has achieved foreign-exchange savings and increased rural employment. An expansion project that would double ethanol production at Triangle to 80 million liters per year is now being evaluated. NOCZIM has estimated it is now saving 19 percent of the foreign exchange spent on regular petrol imports and this has been increasing each of the last four years, reaching $1.5 million a year.

Lessons Learned

• Renewable biomass can often be used as a substitute for nonrenewable fossil fuels.

• Cooperation between government and industry can generate public policy results and private profits.

• Technology transfer can be facilitated by systematic training of local work forces, by adapting existing technology to local conditions, and by cooperation with local vendors of goods and services.

Contact Person

David O. Hall
Kings College London
Campden Hill Road
London W8 7AH, UK

Case 17.3
E.I.D. Parry: Integrated Rural Development

One of the main lessons in the World Commission on Environment and Development's report was that projects to achieve sustainable development need to be implemented in an integrated way, involving the full participation of all the stakeholders who have an interest in the projects. Concern for the environment or economic performance alone is not sufficient. Similarly, companies will be better at developing cleaner products and processes with the active support of their employees, the government, and the local community.

There is also increasing evidence that integrated solutions can have a multiplier effect, increasing and spreading the benefits of change. Taking advantage of these multiplier effects is crucial for industry in developing countries, often faced with pressing economic problems of survival. Macroeconomic uncertainty, excessive government intervention, a lack of financial resources, and poor access to best practice and clean technologies act as major drawbacks to sustainable industrial development in these countries. Nevertheless, broad-based efforts, founded on shared goals, can overcome these obstacles to sustainable development.

This was what E.I.D. Parry, based in southern India, found when it embarked on a restructuring program to ensure the viability of its cane sugar operations. Involving the local farmers who provide the raw cane helped to turn the company around from a situation of heavy debt to one of profitability. Although the company's primary objectives were economic, in the process the social and environmental conditions of the local community have been improved.

E.I.D. Parry (India) Limited is a 200-year-old company producing sugar, confectionery, and agricultural inputs, with 1990–1991 sales of Rs2.62 billion ($101 million). But this healthy situation is only a relatively recent phenomenon. Starting in the 1960s, the company went from bad to worse. Throughout the 1970s no dividend was declared for the shareholders. By the early 1980s the company had reached such a weak situation that it had to be taken over by the Murugappa Group, a leading business group in southern India. The accumulated losses of the sugar unit had mounted to Rs150 million ($5.8 million). Straightforward survival of the operations at Nellikuppam in Tamil Nadu was the order of the day for the new management.

Three main issues faced the company: first, ensuring an increased and guaranteed supply of sugarcane; second, improving the efficiency of the sugar processing operations through technological modernization; and third, ensuring that its environmental performance met community expectations.

The first part of Parry's strategy was to make sure it would obtain sufficient supplies of raw cane from the local farming community so that it could operate at full capacity. To do this the company needed to convince approximately 10,000 farmers who supplied it with raw materials that investments in increased cane production would find a ready market at Parry's. So in early 1986 the company put together a development program with the assistance of government and other agencies. It was implemented through finely tuned teamwork under the leadership of P. Chandrasekaran, then vice-president of the sugar unit and the current president of the company. In the words of Umesh Rao, who has been associated with the sugar unit for three decades and who is now its general manager, "The phenomenal turnaround of the sugar unit was achieved by our commitment to regain the confidence of the farmer."

Motivating the farmers to expand sugarcane cultivation required persistent efforts from the company. Parry collaborated with the sugarcane research institute to develop hybrid varieties that are resistant to drought and pests. Nurseries were established for high-yielding and pest-resistant varieties. Advice was given to the farmers on soil analysis, as well as on optimum fertilizer and pesticide use. Local banking institutions were harnessed to provide credit facilities. In addition, the company gave financial incentives for early planting and harvesting, amounting to Rs15 million ($579,000) in 1990. Financial assistance was also offered in the form of loans for sinking deep bore wells and installing pumps to promote the spread of much-needed irrigation. Finally, the company helped minimize the risk for the farmers by providing insurance cover against crop failure.

The results have been impressive: between 1986 and 1990 the area under cultivation increased from 6,280 to 8,000 hectares, the sugar yield per hectare grew from 1.07 to 1.64 metric tons, and the sugar recovery rate climbed from 8.7 percent to 9.2 percent. But these results were not achieved easily. It required a lot of persuasion and education before the cane growers realized the need for a change in cultivation practices. For example, the farmers had traditionally concentrated on certain varieties that were high yielding in terms of sugarcane but poor in terms of the actual sugar recovered. As a result, the company introduced a new payment system based on the actual amount of sugar obtained.

Computerization was a major element of Parry's strategy to improve productivity. To recover more sugar, computers were used to analyze the quality and maturity of the cane so it would be harvested at the right time. The cane weighing and payment systems were also automated to ensure accurate calculation and quick payments to farmers. Since most of the farmers use the credit facilities from the banks with the assistance of the company, payments are credited directly to their bank accounts. In addition, the company introduced new, more efficient production technology, resulting in less downtime and improved efficiency. And a comprehensive training program was established to improve worker productivity.

Since Nellikuppam had literally grown around the Parry factory, the company recognized it had a clear responsibility to the community to achieve an excellent environmental record, beyond that required by any government regulations. Rather than installing an end-of-pipe effluent treatment plant to purify wastewater from the factory, the company invested Rs32 million ($1.2 million) in an anaerobic digestion process, which produces biogas. This is then used to fuel the boilers, saving Rs1 million a year. The effluent is further subjected to an aerobic treatment and let out for irrigation. As the biological oxygen demand of the effluent is reduced to close to 200 parts per million, the treated water can be used directly for irrigating crops. The use of treated effluent reduces the cost of cultivation to the farmers by about 30 percent because of the rich nutrients such as nitrogen, phosphates, and potash it contains. The effluent sludge is also applied as a fertilizer.

The combined impact of these initiatives has been quite positive for the company. Between 1986 and 1990, every Rs2 invested by the company yielded Rs3 in return. But the intangible benefits, such as the renewed confidence of the growers and the community, have also been substantial. Parry's managers attribute the enhanced performance of the sugar mill to the harmonious relationship between the growers and the company. In terms of sugar recovery, Parry has risen from a position of 27th out of 29 mills in Tamil Nadu in 1985 to 10th out of 32 mills in 1990. The company hopes to reach the top in the next few years.

Lessons Learned

• Industry has a vital role to play in the renaissance of rural life and can itself benefit from close cooperation with the local community.

• When a company helps its suppliers upgrade their inputs, both the company and the suppliers can benefit from the efficiency program.

• Farmers can improve their output and profitability if they are given sound information, appropriate technology, and reasonable financing.

Contact Person

Rajat Nandi
Senior Director
Confederation of Engineering Industry
23-26 Institutional Area, Lodi Road
New Delhi 110-003, India

Case 17.4
Aracruz Celulose: Sustainable Forestry and Pulp Production

Long before deforestation of the Amazon became an international issue, the coastal forestland of Brazil was already being subjected to unsustainable development. Forests in Espírito Santo and Bahia—regions north of Rio de Janeiro still noted for their coffee plantations—were being exploited, deforested, and ultimately abandoned. Unfortunately, once the trees were cut, they were not replaced.

By the late 1960s, farmers, charcoal makers, loggers, and livestock raisers had stripped and burned vast tracts of woodlands. Virtually no replanting or reforestation took place. Small-scale farmers tried to plant crops on the deforested lands, but for the most part they failed. Coffee plantations, which proved to be only marginally more successful, were abandoned altogether when world market prices for coffee plummeted. By 1967, large areas of Brazil's southeastern coast suffered from severe soil erosion and depletion problems. In addition, environmental failure was leading to mounting unemployment and poverty.

The situation was not irreversible, however. For the Brazilian company Aracruz Celulose, the combination of entrepreneurial initiative and government financial support made reforestation of the region with plantations of fast-growing eucalyptus an inviting commercial prospect. In the process, the company was able to regenerate both the local economy and environment.

Aracruz Celulose is one of Brazil's leading companies, with 1991 sales of some $300 million. It specializes in kraft pulp for making paper. During the 1960s, the company began to assess the prospects for using degraded lands in Espírito Santo as a location for new plantations of eucalyptus as the basis of a wood pulp business. An extensive research and experimentation program began to identify the most appropriate species for the local environment. Forestry experts confirmed that the area was ideal for eucalyptus plantations. The company then developed an integrated development plan for the designated area of more than 203,000 hectares in Espírito Santo and Bahia. The plan involved substantial research and development into the most appropriate strains of eucalyptus, the establishment of a state-of-the-art pulp mill, a port to distribute the final product, and an extensive social infrastructure for the local community. Total capital costs of the project from its inception in the 1960s exceeded $700 million ($1.3 billion at current values).

The first plantations were established in 1967 using eucalyptus seeds from the state of São Paulo. Later, studies of local climatic and geographical conditions enabled Aracruz's forestry research teams to select seeds from other regions, such as Australia, Timor, Rhodesia, South Africa, Indonesia, and Papua New Guinea, that were more suited to the conditions in Bahia and Espírito Santo. These seeds were planted at different sites and used in a tree improvement program that generated fully adapted eucalyptus plantations. Yields increased from 25 cubic meters of wood per hectare each year to 35 cubic meters.

A major breakthrough was achieved in 1979 when the company developed a method of vegetative propagation based on cloning techniques. This made it possible to grow trees that have uniform wood quality, greatly simplifying the pulping process. These strains were also fully adapted to the local environment, making them disease-resistant.

In 1984, the Marcus Wallenberg Foundation of Sweden awarded its annual prize to the Aracruz Forestry Research Team for its forest development program and research on vegetative propagation.

It takes only seven years for the cloned eucalyptus trees to grow to maturity—a short time compared with the long growth cycle of most other trees. Lands that had been depleted of native trees were replanted with this fast-growing species. The company decided to conserve the remaining 27 percent of the region still covered with the original forest. This means that there is approximately one hectare of forest with native species for every 2.4 hectares of eucalyptus. In addition, 1.5 million native trees have been planted, including 60,000 fruit trees to foster an increase in the bird population. Aracruz also distributes 9 million free eucalyptus seedlings to local farmers every year, which substantially reduces the local pressures to cut down native forests for charcoal or lumber.

Aracruz has always met 100 percent of its pulp needs by using its own managed forests or by purchasing eucalyptus from local growers without including any native trees. In fact, native tropical forests cannot be used efficiently for pulping purposes because the wood is not sufficiently uniform in character. The high yields of the new forests reduce the amount of land required for pulp operations, as well as enabling large, formerly barren tracts to become productive again. Aracruz has also adapted special tractors and wood-handling equipment to prevent soil compaction and to minimize environmental disruption while the forests are harvested and before they can be replanted.

Having ensured a guaranteed supply of raw material, Aracruz established a major kraft pulp mill capable of producing 400,000 metric tons of pulp—subsequently increased to 500,000 tons—which came on stream in 1978. All the chemicals required in the pulping process (sodium chlorate, chlorine, and caustic soda) are manufactured on-site. The Aracruz plant was among the first in Latin America to use the "clean" cell membrane process in its chlorine-alkali plant. This eliminated the earlier, mercury-based process, which produced wastewater with hazardous contaminants. Aracruz has recently decided to invest $100 million to improve brownstock washing, oxygen predelignification, and the elimination of molecular chlorine—all of which reduce pollution. Furthermore, Aracruz has developed a technology to produce totally chlorine-free pulp and has already started industrial trial runs. In 1991, Aracruz inaugurated a second kraft pulp mill, doubling total capacity to more than 1 million metric tons per year. Both mills comply with the most stringent regulations in operation in Europe and North America. For example, wastewater from the plant receives primary and secondary treatment before being discharged through an outfall into the sea, 1,700 meters away from the coast.

The company has developed a complex industrial ecosystem whereby it is not only assured of a guaranteed supply of environmentally sustainable eucalyptus but is also almost self-sufficient in terms of energy. The use of industrial wastes and bark means that biomass now provides more than 90 percent of total energy requirements. The new pulp mill includes a special unit for the thermogeneration of electricity from wastes and bark, and will reduce electricity purchases even further.

Aracruz now employs 7,500 workers. Sixty percent of its employees are natives of the state of Espírito Santo, while 30 percent come from the neighboring states of Minas Gerais and Bahia. The intensive use of local workers has helped diminish the migration flow to overcrowded Brazilian cities. Aracruz's concern for its workers and the local community has led to the construction of recreational centers, elementary schools, vocational training centers, and health care clinics, costing more than $15 million.

The reforestation program has been very successful. Soil depletion and erosion have virtually disappeared. The introduction of eucalyptus has contributed to the regeneration of the soil. To guard against overexploitation, the company continually monitors soil fertility, adding nutrients where necessary. Concerns about the environmental impacts of the eucalyptus plantations have proved to be unfounded. Careful land use planning has meant that the plantations have not invaded the remaining natural forests, as feared. Aracruz has also taken steps to counter the threat of its cloned eucalyptus being ravaged by pests or diseases. The coexistence of eucalyptus and indigenous forests provides a well-balanced environment that controls potential pests and diseases. Furthermore, Aracruz has developed more than 100 different eucalyptus clones and as a matter of company policy has decided never to plant more than 26 hectares with any one clone.

The company does admit, however, that monoculture plantations do not provide as diversified a habitat as indigenous forests. But Aracruz contends that the comparison between eucalyptus and indigenous forests is not relevant in this case, because the eucalyptus did not replace indigenous forests but instead a thoroughly degraded environment with little or no biodiversity. In fact, because of their fast growth and high yields, eucalyptus plantations can help reduce the pressure for logging native forests for wood products. As Erling Lorentzen, chairman of Aracruz, says, "Eucalyptus is a complement to and not a substitute for tropical forests."

Lessons Learned

• Devastated forest areas can offer business opportunities for sustainable forestry.
• Enlightened environmental and social stewardship can be combined with corporate profitability.

Contact Person

Carlos Roxo
General Manager of Environmental and Public Affairs
Aracruz Celulose S.A.
Rua Lauro Muller 116-40 Andar
22290 Rio de Janeiro, Brazil

Case 17.5
ALCOA: Sustainable Mining in the Jarrah Forest

Australia is a country especially well endowed with minerals. In particular, the southwest corner of the state of Western Australia is rich in bauxite, gold, tin, coal, and other minerals. Mining and mineral processing generate A\$3.5 billion (\$2.6 billion) a year. In the 1970s, tensions between development and conservation interests became particularly intense in areas with unique ecosystems, such as the rare eucalyptus forest in Western Australia. Mining and extracting the valuable wealth-producing resources in this area require careful management of the area's water resources, the jarrah forest itself, which is already extensively affected by a microscopic fungal disease, and the species that live in the forest.

In the 1960s and early 1970s, serious concerns were expressed over the ability of mining companies to restore lands and forests after extraction of the resource. As early as 1976 the Australian government had already set aside a large area of jarrah forest that was largely unaffected by disease. The resources in this and other areas could not be developed unless the forestry and mining industries could demonstrate that such operations were compatible with survival of the indigenous forests. The mining industry recognized that its future would be determined by its ability to satisfy the genuine conservation concerns of the public and regulatory authorities.

The major mining companies operating in this area include ALCOA, AMC Mineral Sands, Cable Sands, Griffin Coal, Western Collieries, Westralian Sands, and Worsley Alumina. Each has made significant contributions to managing the impacts of their operations on the forest ecosystem and other land uses. In particular, ALCOA's program of rehabilitation of its bauxite operations serves as an example of sustainable development in the mining industry.

ALCOA, a U.S.-based aluminum producer, has developed a comprehensive approach to bauxite mining that takes in all stages of the life cycle of a mine: planning, operations, and rehabilitation. Before extraction at a new site, 5- and 10-year mining plans, compiled in conjunction with government agencies, are prepared and take into account all relevant environmental considerations. A rigorous examination of drainage patterns and existing water quality leads to assessment of the likely effects of mining on the proposed site's hydrology and the most effective ways of countering them. Flora and fauna baseline surveys establish existing ecosystem characteristics, and form a basis for determining site-specific rehabilitation objectives. About 450 hectares per year are surface mined to an average depth of about four meters and then rehabilitated.

To address the special conditions of the jarrah forest, ALCOA conducts dieback disease surveys with government officials to establish boundaries between healthy forest and fungus-infected areas. Once areas affected are identified, appropriate environmental management procedures are undertaken to minimize the risk of infected soil being carried into healthy forest. These measures are stringently applied to all aspects of operations, from initial ground reconnaissance to final rehabilitation.

In the earliest rehabilitation activities, mine sites were revegetated as timber plantations—first pines, then eucalyptus. The jarrah dieback disease precluded any thought of reintroducing jarrah. Today, however, sophisticated methods of soil and seed preparation, a better understanding of the disease, a move away from transplanting seedlings in favor of direct seeding, and research into cloning technology have made it possible to reestablish jarrah forest communities.

ALCOA's rehabilitation efforts have also involved innovative ways of dealing with problems associated with the stockpiling of the topsoil above the bauxite. Prior to 1976, topsoil was collected alongside the pit from which it had been taken and then, at the completion of mining in that pit—up to three years later—respread during rehabilitation. Subsequent trials demonstrated that better natural regeneration is obtained by moving freshly stripped topsoil directly to a rehabilitation site. This practice is now carried out in all areas except those where excessive haul distances or movement of dieback-infected soil make the practice undesirable. The company has also improved the establishment and stability of trees planted in mined-out pits by breaking up the compacted soil to a depth of 1.8 meters, using a specially designed ripping tine.

The environmental knowledge gained by ALCOA directly and through the company's association with other research institutions is clearly able to help create sustainable development by others in Western Australia. ALCOA has sponsored a postgraduate program in Land Management and Rehabilitation at Ballarat University. The company also actively encourages public tours of its mine sites and freely provides relevant information about its activities on request. Between 25,000 and 28,000 people tour ALCOA mine sites each year.

ALCOA aims to diffuse its expertise as widely as possible. The benefits of such an open approach are many, according to Lauchlan McIntosh, executive director of the Australian Mining Industry Council: "Shared within the industry, experience gained through such work helps accelerate the adoption and continuous improvement of sustainable practices. Shared with the wider community, it will further ease the burden of development on our planet. It will also help take the mining industry to a position of recognized leadership in sustainable development."

Given that ALCOA's ultimate rehabilitation objective is to reestablish jarrah forest communities, one of the best measurement tools is a direct comparison of post- and premining biodiversity. Current monitoring data show that the number of plant species recorded in nine-month-old rehabilitated sites is nearly 80 percent of the number recorded in sites from which topsoil was obtained. Ninety percent of bird species, most mammal species, and 78 percent of reptile species use the revegetated sites for feeding, shelter, or breeding within 5–10 years. Already, four species on Western Australia's rare fauna list have been identified in these areas. It is anticipated that as the revegetated sites age and blend with the surrounding natural forest, they will develop a greater variety of sources of food and shelter, and these percentages will rise even further.

ALCOA's 27 years of mining and rehabilitation work have also been recognized through the company's listing on the U.N. Environment Programme's Global 500 Roll of Honour. ALCOA was the first mining company in the world to receive this award. Recognizing the award on behalf of the government, Sir Ninian Stephen, Australia's Ambassador for the Environment, said that "ALCOA is a splendid example of the way in which a balance between the economic needs of the Australian people and the capacity of the natural environment can be reconciled. ALCOA is showing us what ecologically sustainable development can mean in practice."

Lessons Learned

• Major mining operations in sensitive areas can be conducted in an eco-efficient way when government and an informed public are willing to cooperate in developing high environmental standards and community relations.

• A strong commitment to environmental research and implementation of research findings is necessary to develop the planning and management practices appropriate for a major mining operation in an ecologically sensitive area.

• It is possible to protect sensitive forest areas and reclaim forested land after mining and mineral extraction if management provides proper commitment and gives adequate training, resources, and motivation to its employees.

Contact Person

Lauchlan McIntosh
Executive Director
Australian Mining Industry Council
P.O. Box 363
Mining Industry House
216 Northbound Avenue
Dickson ACT 2602, Australia

Appendix I
Priorities for a Rational
Energy Strategy

Actions with Immediate Effects

Reform energy pricing policies to reflect full environmental cost of energy by removing energy subsidies on the use of polluting sources, such as through:

• abolishing coal subsidies,

• removing company car tax breaks and tax deductions for commuting, and

• putting a value-added tax on electricity and gas.

Actors: Governments, regulators.

Develop and set standards on products and appliances, to make producers and consumers more aware of energy input, using, for example:

• life-cycle energy accounting,

• sectorwide best energy efficiency production standards set by industry associations,

• energy use and life-cycle energy labelling on cars and refrigerators, and

• energy efficiency standards for buildings.

Actors: Industry jointly with governments, with standards set by International Standards Organization or industry associations.

Shift policy mix toward more economic instruments, to reduce carbon dioxide (CO_2) and other emissions in the most effective manner, by:

• tradable permits and

• incentives/disincentives to correct market signals in transition period.

Actors: Governments, U.N. bodies.

Attack waste, through, for example:

• improving power-plant efficiency in developing countries;

• encouraging cogeneration;

• promoting increased recycling of paper, glass, metals, and used oil;

• capturing methane emissions from landfills; and

• reducing gas leaks (as in Soviet pipelines) and gas flaring.

Actors: Governments, industry, consumers.

Improve energy efficiency by closing the gaps between technically and economically possible energy-efficient technologies and their practical achievement. Support may come through:

• sectorwide promotion through industry associations,

• tax incentives or new accounting practices encouraging the replacement of old equipment with more-efficient technology,

• more-open market economies with effective price signals, and

• competition among and privatization of utilities.

Actors: Industry, governments.

Cooperate with Eastern Europe to contain nuclear risks.

Actors: European and other governments, international organizations, industry.

Actions with Medium-Term Effects

Make investments in energy-efficient technologies more attractive, such as:

• financing modernization of the power sector in Eastern Europe and developing countries;

• priming emerging markets for new energy-efficient products, by government purchase schemes for new products, such as refrigerators, housing, and equipment in public projects; and

• providing financial and infrastructure support for introducing energy-efficient products in developing countries.

Actors: Governments, World Bank, U.N. bodies.

Shift energy mix toward more sustainable sources, for example through:

• integrating alternative energy sources into present infrastructure;

• increasing the share of small-scale hydropower projects in developing countries;

• promoting more-efficient use of biomass, especially in developing countries.

Actors: Governments, U.N. bodies, industry.

Accelerate research in promising technologies, such as biomass gasification, clean coal technology, next-generation nuclear power plants, and hydrogen, through:

• joint research programs between industry and governments, and

• refocussing national research programs.

Actors: Governments, research laboratories, and large industry groups.

Improve access to latest energy technology and management expertise in Eastern Europe and developing countries through:

• technology clearinghouses, such as the International Energy Agency's Greenhouse Gas Technologies, or the U.N. Environment Programme's International Cleaner Production Clearing House system; and

• a buildup of training programs and infrastructure for the introduction of new technology.

Actors: Governments.

Actions with Long-Term and Permanent Effects

Increase knowledge of climate change and its impact by supporting research projects on:

• climate change science, and the impact of climate change; and

• economic policy implications and systems analysis.

Actors: Scientific institutions.

Develop more-efficient energy technologies, through research cooperation on:

• fuel cells,

• intrinsically safe nuclear energy,

• photovoltaics and other solar energies,

• coal gasification,

• high-efficiency transport equipment, and

• biomass-based fuels.

Actors: Government and industry research programs.

Develop cost-efficient CO_2 sinks to compensate for emissions, such as:

• domestic and international afforestation programs,

• research in CO_2 absorption, and

• improved forest management and biomass handling.

Actors: Governments, U.N. bodies, and industry, through incentives and voluntary initiatives.

Develop new urban and regional infrastructures that limit CO_2 emissions, through, for example:

• investment in new transport systems, and

• new guidelines for urban and regional planning.

Actors: Governments.

Evaluate the introduction of an "insurance premium" on the emissions of pollution even when we do not know the full environmental cost, such as through:

• carbon content levies.

Actors: Organisation for Economic Co-operation and Development, policy research institutes, governments, industry.

Promote changes in life-styles, by encouraging public transport and more environmentally sustainable consumption patterns, through:

• better information on energy intensity of products, and

• education and public awareness programs.

Actors: Governments, U.N. bodies, industry.

Specific Initiatives by Industry

• Assess energy efficiency investments at the lowest discount rate applicable to any investment.

• For utilities, set up "best practice" systems to aid developing-country utilities.

• Set up joint international efforts to tackle major projects, such as Soviet gas pipeline leaks.

• Take the lead in energy use labelling on products and processes.

• Make staff available to help East European and developing-country companies with energy efficiency and audit efforts.

• Initiate with governments long-term energy strategies consistent at the national, regional, and global level, such as the Japanese Action Programme to Arrest Global Warming ("New Earth 21").

Appendix II
Acknowledgments

Since it began its work in late 1990, the Business Council for Sustainable Development has received advice, counsel, and data from hundreds of people and organizations. Many of these were not associated directly with the Council or its members, but gave freely of their scarce and valuable time, as they realized the importance of business's role in shaping a sustainable future for the planet and its people. We are deeply indebted to all who have supported our work, and regret that we cannot list them all here:

Mohammed El-Ashry, World Bank, United States
Richard Blackhurst, GATT, Switzerland
Silvio Borner, University of Basel, Switzerland
John Deary, Confederation of Zimbabwe Industries, Zimbabwe
Charles de Haes, WWF, Switzerland
Arthur Dunkel, GATT, Switzerland
Bruno Fritsch, ETH, Switzerland
Daniel Goeudevert, Volkswagen, Germany
Ray Goldberg, Harvard School of Business Administration, United States
José Goldemberg, Minister of Education, Brazil
Luis Gomez-Echeverri, UNDP, United States
Martin Holdgate, IUCN, Switzerland
Nay Htun, UNCED, Switzerland
Ashok Khosla, Development Alternatives, India
Pedro-Pablo Kuczynski, First Boston, United States/Peru
Martin Lees, International Consultant, France
Nicholas Livingston, Gautier, Salis Finance, Switzerland
Eduardo Lizano, CEF, Costa Rica
Christian Lutz, GDI, Switzerland

L. Riitho Ndungi, The Kenya Association of Manufacturers, Kenya
Robert L. Paarlberg, Center for International Affairs, Harvard University, United States
David Pearce, University College, United Kingdom
Guy Pfefferman, IFC, United States
David de Pury, ABB, Switzerland
Juan Rada, IMD, Switzerland
Robert C. Repetto, WRI, United States
Gert Rosenthal, CEPAL, Chile
William Ryrie, IFC, United States
Martyn J. Riddle, IFC, United States
Richard Sandbrook, IIED, United Kingdom
Wolfgang Schürer, Management Service, Switzerland
Katsuo Seiki, MITI, Japan
Hernando de Soto, ILD, Peru
Gus Speth, WRI, United States
Andrew Steer, World Bank, United States
Ernst U. von Weizsäcker, Institute for Climate, Environment and Energy, Germany
Donna W. Wise, WRI, United States

BCSD Liaison Group - Associates and Assistants of BCSD Members

Wendy Ackerman, Pick'n Pay Stores, South Africa
Robert M. Aiken, E.I. du Pont de Nemours and Company, United States
Yoshi Amamiya, Oji Paper, Japan
Thomas P. d'Aquino, Business Council on National Issues, Canada
Allen H. Aspengren, 3M Europe, Belgium
Patrick R. Atkins, ALCOA, United States
Pedro Bernad Moreno, Tecnología Ambiental, Spain
Robert Bringer, 3M, United States
David T. Buzzelli, Dow Chemical, United States
Gaetano Cecchetti, ENI, Italy
Sarawoot Chayovan, Federation of Thai Industries, Thailand
Ivan Cheret, Lyonnaise des Eaux-Dumez, France
Martin G. Colladay, ConAgra, United States
John R. Dillon, Business Council on National Issues, Canada
Julien Dobongna, Compagnie Financière et Industrielle, Cameroon
Tomo Edagawa, Oji Paper, Japan

Jan M. Edelstein, American International Group, United States
Saburo Eto, Keizai Doyukai, Japan
Prince Lekan Fadina, Equity Securities, Nigeria
Carlos H. Fernández, BHN Multibanco, Bolivia
Raymond Florin, IBS Investment Banking Services, Argentina
Michel Gisiger, Société Générale de Surveillance, Switzerland
Dolores Gregory, Browning Ferris Industries, United States
Eric W. Gustafson, DUMAC, Mexico
Claes G. Hall, Aracruz International, United Kingdom
Masayo Hasegawa, Sasakawa Peace Foundation, Japan
Sekio Higuchi, Nissan Motor, Japan
Jane M. Hutterly, S.C. Johnson & Son, United States
L. Oakley Johnson, American International Group, United States
Utomo Josodirdjo, P.T. Artha Investa Argha, Indonesia
Masao Kadota, Kyocera, Japan
Margaret G. Kerr, Northern Telecom, Canada
Yasuyuki Koie, Tosoh, Japan
Shinji Kurata, Tosoh, Japan
Klaus M. Leisinger, Ciba-Geigy, Switzerland
Michael Le Q. Herbert, Shell International Petroleum, United Kingdom
Rolf Marstrander, Hydro Aluminium, Norway
Lauchlan McIntosh, Australian Mining Industry Council, Australia
Stefan Melesko, Axel Johnson, Sweden
Nicola Mongelli, ENI, Italy
Kyosuke Mori, Mitsubishi, Japan
William J. Mulligan, Chevron, United States
Tunku Nadzaruddin ibni Tuanku Ja'afar, Antah Holdings Berhad, Malaysia
Rajat Nandi, Confederation of Indian Industry, India
Heinz Noesler, Henkel, Germany
Anders Norlander, Axel Johnson, Sweden
Catharina Nystedt-Ringborg, ABB Fläkt, Sweden
Carlos Alberto de Oliveira Roxo, Aracruz Celulose, Brazil
Dhira Phantumvanit, Thailand Development Research Institute, Thailand
Michel Potvin, Nissan European Technology, Belgium
Philip K. Rees, John Laing, United Kingdom
Haakon Sandvold, Norsk Hydro, Norway
Frank Schakau, Volkswagen, Germany
Anne Seba, Pick'n Pay Stores, South Africa
Ulrich Steger, Volkswagen, Germany

W. Ross Stevens III, E.I. du Pont de Nemours and Company, United States
Björn Stigson, ABB Fläkt, Sweden
François Thiesse, L'Air Liquide, France
Sergio Verdugo, CAP, Chile
Ben Woodhouse, Dow Chemical, United States
Makoto Yoshida, Nippon Steel, Japan

BCSD Staff and Consultants

* J. Hugh Faulkner, Executive Director, Canada
  Stephen Bass, United Kingdom
* Frank W. Bosshardt, Switzerland
* Ernst A. Brugger, Switzerland
  Tim Crabtree, United Kingdom
  Márcio Fortes, Brazil
  Albert E. Fry, United States
  Barbara Gorsler, Germany
  David Harris, United Kingdom
  Erich Heini, Switzerland
  Matthew J. Kiernan, Canada
  Serge de Klebnikoff, France
  Olav Ketilsson, Norway
  Peter Knight, United Kingdom
  David Nilsson, Sweden
  Nick Robins, United Kingdom
  Trevor Russel, United Kingdom
  Linda Starke, United States
  Hans O. Staub, Switzerland
  Francisco Szekely, Mexico
* Lloyd Timberlake, United States
  David Tinnin, United States
  Alex Trisoglio, United Kingdom
  Jan-Olaf Willums, Norway
  Mark G. Wilson, United Kingdom
  Hugh Wynne-Edwards, Canada
  *Members of the Editorial Committee

Organizations

Asociación de Industriales Latinoamericanos, Latin America

Asociación Iberoamericana de Camaras de Comércio, Latin America

Confederation of Indian Industry (CII), India

Confederation of Zimbabwe Industries, Zimbabwe

Confederação Nacional da Agricultura, Brazil

Confederación Latinoamericana de la Mediana y Pequeña Industria, Latin America

Conference Board of Canada, Canada

DRT International, Canada

Federación de Entidades Privadas de Centroamérica y Panamá, Costa Rica

Fundação Getulio Vargas, Centro de Economia Mundial, Brazil

General Agreement on Tariffs and Trade (GATT) Secretariat, Switzerland

International Chamber of Commerce (ICC), France

International Council for Development (ICD), United States

International Environmental Bureau (IEB), Norway

International Finance Corporation (IFC), United States

International Institute for Environment and Development (IIED), United Kingdom

International Institute for Sustainable Development (IISD), Canada

International Institute for Management Development (IMD), Switzerland

International Trade Center (ITC), Switzerland

International Union for Conservation of Nature and Natural Resources (IUCN), Switzerland

Keidanren, Japan

Keizai Doyukai, Japan

Pro Rio 92, Brazil

Sasakawa Peace Foundation, Japan

United Nations Development Programme (UNDP), United States

United Nations Environment Programme (UNEP), Kenya

UNEP—Industry and Environment Office (UNEP-IEO), France

World Bank, United States

World Economic Forum, Switzerland

World Resources Institute (WRI), United States

World Wide Fund for Nature (WWF), Switzerland

Notes

Chapter 1

1. U.N. Environment Programme, *Environmental Data Report 1989/90* (Oxford: Basil Blackwell Ltd., 1989).

2. World Bank, *World Development Report 1990* (Washington, D.C.: 1990).

3. Havelock Brewster, "Third World's Prospects in the World Economy in the 1990s," *Third World Economics*, July 16–31, 1991.

4. Maurice Strong, Speech to UNEP-UK Committee, London, April 1991.

5. International Chamber of Commerce (ICC), *WICEM II: Conference Report and Background Papers* (Paris: 1991).

6. Michael Porter, "Green Competitiveness," *New York Times*, June 5, 1991.

7. World Commission on Environment and Development (WCED), *Our Common Future* (Oxford: Oxford University Press, 1987).

8. Ibid.

9. International Union for Conservation of Nature and Natural Resources et al., *World Conservation Strategy* (Gland, Switzerland: 1980).

10. "Economic Declaration," Summit of the Arch, Paris, July 16, 1989.

11. ICC, *WICEM II*.

12. Keidanren, "Keidanren Global Environmental Charter," Tokyo, April 23, 1991.

13. Malaysian Environmental Quality Council, "Malaysia's Corporate Environmental Policy," 1991.

14. Confederation of Engineering Industry, "Environment Code for Industry," New Delhi, 1991.

15. Chemical Manufacturers Association, "Responsible Care Brochure," Washington, D.C., 1991.

16. Ibid.

17. Robert Goodland et al. (eds.), *Environmentally Sustainable Economic Development: Building on Brundtland* (Paris: UNESCO, 1991).

18. WCED, *Our Common Future*.

19. Donella Meadows et al., *The Limits to Growth* (New York: Universe Books, 1972).

20. Organisation for Economic Co-operation and Development, *The State of the Environment* (Paris: 1991).

Chapter 2

1. W. Ross Stevens III, E.I. du Pont de Nemours and Company, personal communication, June 1991.

2. "Making Effective Use of Economic Instruments and Market Incentives," in "Integration of Environment and Development in Decision Making," Document A/CONF.151/PC/1OO/Add.8, Preparatory Committee for the U.N. Conference on Environment and Development, Fourth Session, New York, March 2–April 3, 1992.

3. Jeffrey Leonard, BCSD Workshop on Economic Instruments, Bedford, U.K., July 12-14, 1991.

4. T. Tietenberg, "Economic Instruments for Environmental Regulation," *Oxford Review of Economic Policy*, Vol. 6, No. 1, 1990.

5. Robert Stavins (ed.) , *Project 88—Round Two* (Washington, D.C.: 1991).

6. Ibid.

7. M. Pearson and S. Smith, "Taxation and Environmental Policy: Some Initial Evidence," IFS Commentary No. 19, Institute for Fiscal Studies, London, 1990.

8. Organisation for Economic Co-operation and Development, "Recent Developments in the Use of Economic Instruments," Environment Monographs No. 41, Paris, 1991.

9. Ibid.

10. Theodore Panayotou, "Economic Instruments for Hazardous Waste Management: The Proposed Industrial Environment Fund of Thailand," paper presented to BCSD Workshop on Economic Instruments.

11. Harry F. Campbell, "Future Directions: Economic Instruments and Marine Resources," paper presented to BCSD Workshop on Economic Instruments.

12. David Pearce et al., *Blueprint for a Green Economy* (London: Earthscan Publications Ltd., 1990).

13. Robert Repetto et al., *Wasting Assets: Natural Resources in the National Income Accounts* (Washington, D.C.: World Resources Institute, 1989).

14. David Pearce and Karl-Göran Mäler, "Environmental Economics and the Developing World," *Ambio*, April 1991.

Chapter 3

1. *Biomass Users Network*, July/August 1991.

2. BCSD meeting at International Energy Agency (IEA), Paris, May 23, 1991.

3. Intergovernmental Panel on Climate Change (IPCC), *Formulation of Response Strategies* (Geneva: World Meteorological Organization, 1990); BCSD at IEA.

4. Eric Hirst, "Electricity: Getting More with Less," *Technology Review*, July 1990.

5. IPCC, *Formulation of Response Strategies*.

6. Frances Cairncross, "Energy and the Environment Survey," *The Economist*, August 31, 1991.

7. IPCC, *Formulation of Response Strategies*.

8. Ibid.

9. Peter Martyn et al., "Aluminum and Energy," *Industry and Environment*, April/May/June 1990.

10. Roger Rainbow, Shell International, London, personal communication.

11. William Fulkerson et al., "Energy from Fossil Fuels," *Scientific American*, September 1990.

12. *Industry and Environment*, April/May/June 1990.

13. Carl J. Weinberg and Robert H. Williams, "Energy from the Sun," *Scientific American*, September 1990.

14. Björn Stigson, Chief Executive Office, ABB Fläkt, speech.

15. *Biomass Users Network*.

16. See case 17.2 in chapter 17.

17. B.V. Chitnis, Tata Consulting Engineers, Bombay, India, personal communication at BCSD meeting, Geneva, June 19, 1991.

18. Presentation at NGO Forum, Preparatory Committee for the U.N. Conference on Environment and Development, Third Session, Geneva, August 1991.

19. Chitnis, personal communication.

Chapter 4

1. International Finance Corporation, *Emerging Stock Markets Factbook 1991* (Washington, D.C.: 1991).

2. International Economics Department, World Bank, Washington, D.C., private communication, December 1991.

3. "The Surprising Emergence of Distant Shares," *The Economist*, November 16, 1991.

4. William Ryrie, "Capitalism and Third World Development," *International Economic Insights*, May/June 1991.

5. Ibid.

6. World Commission on Environment and Development, *Our Common Future* (Oxford: Oxford University Press, 1987).

7. Organisation for Economic Co-operation and Development (OECD), *External Debt Statistics* (Paris: 1991).

8. Investor Responsibility Research Center (IRRC), "Investor's Environmental Report, Summer 1991," Washington, D.C., 1991.

9. G. Van Velsor Wolf Jr., "EPA's Lender Liability Rule: No Surprises But More Work Needed," Environmental Law Reporter, January 1991.

10. Peter Drucker, "Reckoning with the Pension Fund Revolution," *Harvard Business Review*, March/April 1991.

11. Anders Wijkman and Lloyd Timberlake, *Natural Disasters: Acts of God or Acts of Man?* (London: Earthscan Publications Ltd., 1984).

12. World Bank, *The World Bank and the Environment, A Progress Report, Fiscal 1991* (Washington, D.C.: 1991).

13. OECD, *Development Co-operation—Efforts and Policies of the Members of the Development Assistance Committee, 1991* (Paris: 1991).

14. Frances Cairncross, *Costing the Earth* (London: The Economist Books, 1991).

15. IRRC, "Investor' s Environmental Report, Spring 1991," Washington, D.C., 1991.

16. "The Surprising Emergence of Distant Shares."

Chapter 5

1. S. Fardoust and A. Dhareshwar, *A Long-Term Outlook for the World Economy; Issues and Projections for the 1990s* (Washington, D.C.: World Bank, 1990).

2. Winfred Ruigrok, "Paradigm Crisis in International Trade Theory," *Journal of World Trade*, February 1991.

3. "Business This Week," *The Economist*, July 27–August 2 , 1991.

4. Robert Reich, "Who is Them?" *Harvard Business Review*, March/April 1991.

5. Organisation for Economic Co-operation and Development (OECD), *The State of the Environment* (Paris: 1991).

6. "United States—Restrictions on Tuna," GATT Arbitration Panel Report, Geneva, September 3, 1991.

7. Hilary French, "Strengthening Global Environmental Governance, " in Lester R. Brown et al., *State of the World 1992* (New York: W.W. Norton & Co., 1992).

8. GATT Information Office, Geneva, undated.

9. "Agreement on Technical Barriers to Trade," GATT, Geneva, April 12, 1979.

10. U.N. Development Programme, *Human Development Report 1991* (New York: Oxford University Press, 1991).

11. Sam Laird and Alexander Yeats, "Trends in Nontariff Barriers of Developed Countries: 1966 to 1986," World Bank PPR Working Paper 137, Washington, D.C., December 1988.

12. OECD, *Agricultural Policies, Markets and Trade: Monitoring and Outlook 1991* (Paris: 1991).

13. Quoted in Keith Bradsher, "In Mexico, Fears of Free Trade Melt," *New York Times*, September 22, 1991.

14. U.N. Centre on Transnational Corporations, "The Triad in Foreign Direct Investment," New York, 1991.

15. OECD, *Development Co-operation—Efforts and Policies of the Members of the Development Assistance Committee, 1991* (Paris: 1991).

16. T. Haavelmo and S. Hansen, "On the Strategy of Trying to Reduce Economic Inequality by Expanding the Scale of Human Activity," in Robert Goodland et al. (eds.), *Environmentally Sustainable Economic Development: Building on Brundtland* (Paris: UNESCO, 1991).

17. Press Release, OECD, Paris, December 5, 1989.

Chapter 6

1. See case 15.1 in chapter 15.

2. *Solidarity* (United Auto Workers of America), March/April 1991.

3. Caroline Sargent, *The Khun Song Plantation Project* (London: International Institute for Environment and Development, 1990).

4. See case 12.4 in chapter 12.

5. Ministry of Housing, Physical Planning and Environment, *National Environmental Policy Plan Plus* (The Hague: 1990).

6. William Reilly, "Free Enterprise—Technology—and a Cleaner Environment," *Business Week*, December 30, 1991.

7. *Newsweek*, November 19, 1990.

8. See case 12.3 in chapter 12.

9. See case 11.1 in chapter 11.

10. Arthur D. Little Company, Wiesbaden, Germany, personal communication.

11. See case 13.4 in chapter 13.

12. See case 15.4 in chapter 15.

13. Deloitte & Touche, survey commissioned by BCSD and the International Institute for Sustainable Development, Toronto, 1991.

14. International Chamber of Commerce, "Position Paper on Environmental Auditing," Paris, 1989.

Chapter 7

1. Organisation for Economic Co-operation and Development (OECD), *Technology in a Changing World* (Paris: 1991).

2. Ibid.

3. Nick Robins, *Managing the Environment: The Greening of European Business* (London: Business International, 1990).

4. John R. Ehrenfeld, "Technology and the Environment: A Map or Mobius Strip," quoted in George Heaton et al., *Transforming Technology* (Washington, D.C.: World Resources Institute, 1991).

5. Joel S. Hirschhorn and Kirsten U. Oldenberg, *Prosperity Without Pollution* (New York: Van Nostrand Reinhold, 1991).

6. Clive Cookson, "The New Way to Make a Clean Break," *Financial Times*, April 10, 1991.

7. Lucien Chabason, Director, Plan Vert, speech at World Industry Conference on Environmental Management II, Rotterdam, April 1991.

8. D.B. Redington, "Corporate Transition to Multimedia Waste Reduction," speech at EPA/IACT International Conference on Pollution Prevention, Washington, D.C., June 1990.

9. OECD, *The State of the Environment* (Paris: 1991).

10. Sandra Vandermerwe and Michael Oliff, "Corporate Challenges for an Age of Reconsumption," *Columbia Journal of World Business*, Fall 1991.

11. Quoted in Barry Commoner, *Making Peace with the Planet* (New York: Pantheon Books, 1990).

12. Robert U. Ayres, *Eco-Restructuring: Managing the Transition to an Ecologically Sustainable Economy* (Laxenburg, Austria: International Institute for Applied Systems Analysis, 1991).

13. See case 11.2 in chapter 11.

14. PRISMA, *Choosing for Prevention is Winning* (Gravenhage, Netherlands: Ministry of Economic Affairs, 1991).

15. Ibid.

16. Hirschhorn and Oldenberg, *Prosperity Without Pollution*.

17. Prasad Modak, *Environmental Aspects of the Textile Industry* (Paris: U.N. Environment Programme—Industry and Environment Office, 1991).

18. Monsanto in "Managing Earth's Resources," *Business Week*, June 18, 1990; for 3M, see case 11.2 in chapter 11.

19. Quoted in Robins, *Managing the Environment: The Greening of European Business*.

20. Ibid.

21. D. Huisingh et al., *Proven Profits from Pollution Prevention* (Washington, D.C.: Institute for Local Self-Reliance, 1986).

22. "Managing Earth's Resources," *Business Week*.

23. Modak, *Environmental Aspects of the Textile Industry*.

24. Huisingh et al., *Proven Profits from Pollution Prevention*.

25. Robert A. Frosch and Nicholas E. Gallopoulos, "Strategies for Manufacturing," *Scientific American*, September 1989.

26. Ibid.

27. See case 11.3 in chapter 11.

28. "Willow Dust," Industry and Environment Office, U.N. Environment Programme, Paris, undated.

29. *Environment Business*, March 13, 1991.

30. U.S. Congress, Office of Technology Assessment, *Serious Reduction of Hazardous Waste* (Washington D.C.: U.S. Government Printing Office, 1986).

31. OECD, *Environmental Policy and Technical Change* (Paris: 1985).

32. OECD, "The Promotion and Diffusion of Clean Technologies in Industry,"

OECD Environment Monographs No. 9, Paris, 1987.

33. George Moellenkamp, "Experiences with Environmental Auditing and Industry's Management Attitudes Towards Environmental Issues," at Diagnostics et Audits de l'Environment Conference, Paris, January 1991.

34. Paul R. Wilkinson, "Measuring and Tracking Waste," Du Pont Safety and Environmental Seminars, at the Global Pollution Prevention '91 Conference, Washington, D.C., 1991.

35. Modak, *Environmental Aspects of the Textile Industry*.

36. OECD, *The State of the Environment*.

37. See case 12.6 in chapter 12.

38. Cynthia Pollock, "Realizing Recycling's Potential," in Lester R. Brown et al., *State of the World 1987* (New York: W.W. Norton & Co., 1987).

39. Robert P. Bringer and David M. Benforado, "Pollution Prevention as Corporate Policy: A Look at the 3M Experience," *The Environmental Professional*, Vol. 11, 1989.

40. Adapted from World Wildlife Fund/Conservation Foundation, *Getting at the Source: Strategies for Reducing Municipal Solid Waste* (Washington, D.C.: 1991).

41. See case 16.4 in chapter 16.

42. See case 16.2 in chapter 16.

43. See case 16.6 in chapter 16.

44. Robins, *Managing the Environment: The Greening of European Business*.

45. See case 16.1 in chapter 16.

46. See case 16.3 in chapter 16.

47. OECD, *Eco-labelling in OECD Countries* (Paris: 1991).

48. Nigel Haigh, Institute for European Environmental Policy, London, personal communication, January 17, 1992.

49. Interview with Ulrich Golüke, August 5, 1991.

50. See case 16.7 in chapter 16.

51. Global Industrial and Social Progress Research Institute, "Sustainable Development in the Pacific Rim," Pacific Rim Conference, Tokyo, February 1991.

52. *Economist*, November 30, 1991.

53. Calestous Juma et al., *Industrialisation, Environment and Employment* (Nairobi: African Centre for Technology Studies, 1990).

Chapter 8

1. U.N. Centre on Transnational Corporations, "Transnational Corporations and Manufacturing Exports from Developing Countries," U.N. Publication ST/CTC/101, New York, 1990.

2. U.N. Environment Programme, *Register of International Treaties and Other Agreements in the Field of the Environment* (Nairobi: 1991).

3. Ibid.

4. World Bank, *World Development Report 1991* (New York: Oxford University Press, 1991).

5. Ruth Leger Sivard, *World Military and Social Expenditures*, 1991 (Washington, D.C.: World Priorities, 1991).

6. Lloyd Timberlake, *Africa in Crisis* (London: Earthscan Publications Ltd., 1988).

7. Cherie Hart, "Reversing the Flow: Looking South for Answers," *World Development*, September 1991.

8. Harro Pitkänen, Director, Nordic Environment Finance Corporation, Helsinki, Finland, personal communication, July 3, 1991.

9. Keidanren, "Keidanren Global Environmental Charter," Tokyo, April 23, 1991.

10. See case 13.4 in chapter 13.

11. Economic and Social Council of the United Nations, E/C1O/1991/2.

12. U.N. Development Programme, *Human Development Report 1991* (New York: Oxford University Press, 1991).

13. U.N. Economic Commission for Europe, "ECE Data Bank on East-West Joint Ventures," Geneva, 1991.

14. See case 12.2 in chapter 12.

15. See case 12.5 in chapter 12.

16. See case 11.5 in chapter 11.

17. I.I. Elwan, Manager, Private Sector Financial Operations Cofinancing and Financial Advisory Services, World Bank, Washington, D.C., personal communication, June 5, 1991.

18. See case 13.1 in chapter 13.

Chapter 9

1. Ray A. Goldberg, "Outline for the Business Council for Sustainable Development—Agribusiness and Forestry Policy Study," Harvard University, Cambridge, Mass., July 1991.

2. P.J. Stewart, "The Dubious Case for State Control," *Ceres*, Vol. 18, No. 2, 1985.

3. Norsk Hydro, *Agriculture and Fertilizers* (Porsgrunn, Norway: 1990).

4. World Bank, *Global Food—Resources and Prospects* (Washington, D.C.: 1991).

5. Dennis T. Avery, *Global Food Progress 1991* (Indianapolis, Ind.: Hudson Institute, 1991).

6. U.S. Department of Agriculture, *Our American Land: 1987 Yearbook of Agriculture* (Washington, D.C.: U.S. Government Printing Office, 1987).

7. Luther Tweeten, "The Economics of an Environmentally Sound Agriculture (ESA)," Moline, Ill., mimeographed, 1990.

8. Ibid.

9. U.S. Department of Agriculture, *1989 Yearbook of Agriculture* (Washington, D.C.: U.S. Government Printing Office, 1989).

10. World Bank, *World Development Report 1991* (New York: Oxford University Press, 1991).

11. Avery, *Global Food Progress 1991*.

12. U.N. Food and Agriculture Organization (FAO), *Sustainable Development and Natural Resource Management* (Rome: 1990).

13. FAO, *Agriculture: Toward 2000* (Rome: 1987).

14. Jeffrey Leonard (ed.), *Environment and the Poor: Development Strategies for a Common Agenda* (New Brunswick, N.J.: Transaction Books, for Overseas Development Council, 1989).

15. Alan B. Durning and Holly B. Brough, *Taking Stock: Animal Farming and the Environment* (Washington, D.C.: Worldwatch Institute, 1991).

16. World Health Organization, "Assessment of Mortality and Morbidity Due to Unintentional Pesticide Poisonings," WHO/VBC/86.929, Geneva, 1986.

17. R. Repetto and T. Holmes, "The Role of Population in Resource Depletion," *Population and Development Review*, December 1983.

18. Lester R. Brown et al., *State of the World 1990* (New York: W.W. Norton & Co., 1990).

19. U.N. Economic Commission for Latin America and the Caribbean, *Our Own Agenda* (New York: U.N. Development Programme, 1990).

20. Robert Paarlberg and Michael Lipton, "Changing Missions at the World Bank," *World Policy Journal*, Summer 1991.

21. Avery, *Global Food Progress 1991*.

22. World Bank, *World Development Report 1992* (draft) (Washington, D.C.: November 1991).

23. Agnes Kiss, "The Africa-Wide Biological Control Program for Cassava Mealybug," in A. Kiss and F. Meerman (eds.) , *Integrated Pest Management and African Agriculture*, Technical Paper No. 142 (Washington, D.C.: World Bank, 1991).

24. Charles Benbrook, "Grappling with the Challenges of Sustainability: The Den Bosch Declaration," in *World Food Regulation Review 1991*.

25. Stewart, "The Dubious Case for State Control."

26. Swedish Pulp and Paper Association, "Forests That Never Run Out," *Skogsindustrierna*, September 16, 1991.

27. World Bank, *Forestry Development: A Review of Bank Experience* (Washington, D.C.: 1991).

28. E.O. Asibey, "Development of Private Forest Plantations to Reduce Pressure on Natural Forest in Sub-Saharan Africa," AFTEN Working Paper, 1991.

29. J.E.M. Arnold, "Long-Term Trends in Global Demand for and Supply of Industrial Wood," Oxford Forestry Institute, 1991.

30. E.O. Wilson, *Biodiversity* (Washington, D.C.: National Academy Press, 1988).

31. Edward O. Wilson, "Bedrohung des Artenreichtums," *Spektrum der Wissenschaft*, November 1989.

32. John C. Ryan, "Conserving Biological Diversity," in Lester R. Brown et al., *State of the World 1992* (New York: W.W. Norton & Co., 1992).

33. World Bank, *The Forest Sector*, Policy Paper (Washington, D.C.: 1991).

34. World Resources Institute, *World Resources 1990–91* (New York: Oxford University Press, 1990).

35. Robert Repetto, *The Forest for the Trees? Government Policies and the Misuse of Forest Resources* (Washington, D.C.: World Resources Institute, 1988).

36. World Bank, *The Forest Sector*.

37. Ibid.

38. R.A. Wilson, "Tree Farming: An Environmental Opportunity for the Forest Industry," mimeographed, unpublished, 1990.

39. M.R. de Montalembert, "Key Forest Policy Issues in the Early 1990s," *Unasylva*, No. 166, 1991.

40. Arnold, "Long-Term Trends in Global Demand for and Supply of Industrial Wood."

41. Wilson, *Biodiversity*.

42. Arnold, "Long-Term Trends in Global Demand for and Supply of Industrial Wood."

43. Asibey, "Development of Private Forest Plantations."

44. Arnold, "Long-Term Trends in Global Demand for and Supply of Industrial Wood."

45. World Bank, *The Forest Sector*.

46. Arnold, "Long-Term Trends in Global Demand for and Supply of Industrial Wood."

47. Shell Briefing Service, "Focus on Forestry," Shell International Petroleum Company Ltd., London, October 1990.

48. See case 17.4 in chapter 17.

49. Sandra Postel and John C. Ryan, "Reforming Forestry," in Lester R. Brown et al., *State of the World 1991* (New York: W.W. Norton & Co., 1991).

50. *Turning Values into Value: Ben & Jerry's 1990 Annual Report* (Waterbury, Vt.: Ben & Jerry's Homemade, Inc., 1991).

51. Laura Zinn, "Whales, Human Rights, Rain Forests—And the Heady Smell of Profits," *Business Week*, July 15, 1991.

52. Christopher Joyce, "Prospectors for Tropical Medicine," *New Scientist*, October 19, 1991.

53. Ryan, "Conserving Biological Diversity."

54. Australian Mining Industry Council, *Mining Rehabilitation Handbook* (Dickson, ACT: 1990).

55. Based on Duncan Poore and Jeffrey Sayer, *The Management of Tropical Moist Forest Lands: Ecological Guidelines* (Gland, Switzerland: International Union for Conservation of Nature and Natural Resources, 1987).

56. Ibid.

Chapter 10

1. World Bank, *World Development Report 1991* (New York: Oxford University Press, 1991).

2. Ibid.

3. World Bank, *World Development Report 1992* (draft) (Washington, D.C.: November 1991).

4. World Bank, *World Development Report 1991*; U.N. Development Programme, *Human Development Report 1991* (New York: Oxford University Press, 1991).

5. The South Commission, *The Challenge to the South* (Oxford: Oxford University Press, 1990).

6. Herman E. Daly, "Sustainable Growth: An Impossibility Theorem," *SID Journal*, No. 3/4, 1990.

7. Global Industrial and Social Progress Research Institute, "Sustainable Development in the Pacific Rim," Pacific Rim Conference, Tokyo, February 1991.

8. The New World Dialogue on Environment and Development in the Western Hemisphere, *Compact for a New World* (Washington, D.C.: World Resources Institute, 1991).

9. Population Reference Bureau, "World Population Data Sheet 1991," Washington, D.C. , 1991.

10. UNICEF and U.N. Environment Programme (UNEP), *The State of the Environment 1990: Children and the Environment* (New York: UNICEF, 1990).

11. Nafis Sadik, "Meeting the Challenge of Population Growth and Sustainable Resource Use," manuscript, U.N. Population Fund, New York, April 1991.

12. Klaus Leisinger, "Project Hope," UNICEF, Paris, 1989.

13. Sadik, "Meeting the Challenge of Population Growth and Sustainable Resource Use."

14. Roland Bunch, "Low Input Soil Restoration in Honduras: The Cantarranas Farmer-to-Farmer Extension Programme," Gatekeeper Series No. 23, International Institute for Environment and Development, London, 1990.

15. Robert Wade, *Governing the Market: Economic Theory and the Role of Government in East Asian Industrialization* (Princeton, N.J.: Princeton University Press, 1991).

16. Hernando De Soto, "Coca, Property Rights and Sustainable Development," manuscript, Lima, 1991.

17. World Resources Institute, *World Resources 1990–91* (New York: Oxford University Press, 1990).

18. Gary Kutchner and Pasquale Scandizzo, *The Agricultural Economy of Northeast Brazil* (Baltimore, Md.: Johns Hopkins University Press, 1981).

19. Ernst A. Brugger, "Nachhaltige Entwicklung—eine Herausforderung für die Land—und Forstwirtschaft Lateinamerikas," in W. Hein (ed.), *Umweltorientierte, Entwicklungspolitik* (Hamburg: Deutsches Übersee-Institut, 1991).

20. United Nations, *World Urbanization Prospects 1990* (New York: 1991).

21. World Resources Institute, *World Resources 1990–91*.

22. World Health Organization and UNEP, *Global Pollution and Health* (New Haven, Conn.: Yale University Press, 1987).

23. Organisation for Economic Co-operation and Development, *External Debt Statistics* (Paris: 1991).

24. World Bank, *World Debt Tables 1989–1990* (Washington, D.C.: 1991).

25. World Bank, *The Bretton Woods Agencies and Private Debt* (Washington, D.C.: 1990).

26. David Knox, "Latin American Debt, Facing the Facts," Oxford International Institute, 1990.

27. Inter-American Development Bank, *Economic and Social Progress in Latin America* (Washington, D.C.: 1991).

28. World Bank, *World Debt Tables 1989–1990*.

29. United Nations, *World Economic Survey, 1990* (New York: 1990).

30. Overseas Development Institute, "The Inter-American Development Bank and Changing Policies for Latin America," Briefing Paper, London, 1991.

31. Peter Dogse and Bernd von Droste, "Debt for Nature Exchanges and Biosphere Reserves," UNESCO, Paris, 1990.

32. Michael Porter, "The Competitive Advantage of Nations," *Harvard Business Review*, March/April 1990.

33. Hernando De Soto and Stephan Schmidheiny, *Las Nuevas Reglas del Juego* (Bogotá: Editorial Oreja Negra, 1991).

34. Confederation of Engineering Industry, "The Challenge of a Free Economy," Background Paper, New Delhi, 1991.

35. Quoted in Joseph Contreras, "We Need Our Own Form of Perestroika," *Newsweek*, November 7, 1988.

36. Hernando De Soto, *El Otro Sendero* (Lima, Peru: Instituto Libertad y Democracia, 1986).

37. World Bank, *World Development Report 1988* (New York: Oxford University Press, 1988).

38. Ernst A. Brugger, *The Strategic Importance of Small Enterprise in the Process of Sustainable Development* (Washington, D.C.: in press).

39. Carl Liedholm and Donald C. Mead, "Small-Scale Industry," in Robert J. Berg and Jennifer Seymour Whitaker (eds.), *Strategies for African Development* (Berkeley: University of California Press, 1986).

40. Ernst A. Brugger, "La Función de la Banca Comercial en la Promoción de la Pequeñan Empresa en América Latina," manuscript for 25th Annual Assembly, Federación Latinoamericana de Bancos, Panamá, 1991.

41. See case 14.2 in chapter 14.

42. K.P. Nyati, "Problems of Pollution and Its Control in Small Scale Industries," F. Ebert Foundation, New Delhi, 1988.

43. "Environmental Management of Small and Medium Sized Industries," *Industry and Environment Review*, Vol. 10, No. 2, 1987.

44. International Monetary Fund, *Government Finance Statistics—Yearbook 1990* (Washington, D.C.: 1990).

Chapter 11

1. *National Energy Plan 1977* (Washington, D.C.: U.S. Government Printing Office, 1977).

2. International Chamber of Commerce, "The ICC Guide to Effective Environmental Auditing," ICC Publication No. 483, Paris, 1991.

3. *ENDS*, June 1990.

Chapter 12

1. Quoted in speech by Robert Kennedy, Chief Executive Office, Union Carbide, World Industry Conference on Environmental Management (WICEM) II, Rotterdam, April 1991.

2. Ibid.

3. Speech at WICEM II, April 1991.

Chapter 13

1. *The Nikkei Weekly*, September 21, 1991.

Chapter 15

1. S.K. Jain, "Productivity—Where Do We Stand?" Namma Parisara, Karnataka State Pollution Control Board, 1990.

2. Teesta Setalvad, "Industry Will Have to Start Cleaning Up," *Business India*, May 28–June 10, 1990.

Chapter 17

1. U.S. Congress, Office of Technology Assessment, *Changing By Degrees* (Washington, D.C.: U.S. Government Printing Office, 1991).

2. Ibid.

3. R.R. McKinsey et al., "Engineering and Economic Evaluation of PFBC Power Plants," in American Society of Mechanical Engineers (ASME), *Clean Energy for the World, Vol. 1* (Montreal: 1991).

4. L. Carpenter et al., "PFBC: A Commercially Available Clean Coal Technology," in ASME, *Clean Energy for the World, Vol. 1.*

Index